Paul Zipperer, Herm Schaeffer

The Manufacture of Chocolate and other Cacao Preparations

Paul Zipperer, Herm Schaeffer

The Manufacture of Chocolate and other Cacao Preparations

ISBN/EAN: 9783337364809

Printed in Europe, USA, Canada, Australia, Japan

Cover: Foto ©Andreas Hilbeck / pixelio.de

More available books at **www.hansebooks.com**

PLATE I
The Cacao Tree — Theobroma Cacao, Linné.

Zipperer, Manufacture of Chocolate etc. 3rd edition.
Verlag M. Krayn, Berlin W. 10.

THE
MANUFACTURE

3

OF
CHOCOLATE
AND OTHER CACAO PREPARATIONS

BY
Dr. PAUL ZIPPERER.

THIRD EDITION
REARRANGED, THOROUGHLY REVISED, AND LARGELY REWRITTEN.

EDITOR
DR. PHIL. HERM. SCHAEFFER
FOOD CHEMIST AND MANAGING DIRECTOR.

WITH 132 ILLUSTRATIONS, 21 TABLES AND 3 PLATES.

BERLIN W.
VERLAG VON M. KRAYN.

LONDON		NEW YORK
E. & F. N. SPON Ltd.	1915	SPON & CHAMBERLAIN
PUBLISHERS		PUBLISHERS
57 HAYMARKET.		123-125 LIBERTY STREET.

Preface to the third edition of "The Manufacture of Chocolate" by Dr. Zipperer.

It is now a decade since the appearance of the last edition, and owing to continual delays in the compiling of the present volume, the book has been out of print for several years. These delays ensued because the editor wished to take into account the most recent determinations and decrees of the guilds and various legislative factors connected with the industry; but he was at length forced to the conclusion that notwithstanding the excellent organisation and lofty standing of the branch under consideration, it was useless to wait for anything final and absolute in such a field. Suggestions of possible improvements and indications of blemishes are therefore earnestly invited, in order that they may be duly allowed for in the event of a new edition. — The plan followed by Zipperer has been adopted in the main; a tribute due to its previous success. Yet on the other hand, the arrangement of the book has undergone some alteration, and is, at least in the editor's opinion, a perceptible improvement. — All scientific, industrial and technical progress has been treated as fully as possible, the economic part in particular having been diligently recast.

It would, of course, have been impossible for the editor to write all these chapters without external aid, his knowledge of the respective branches being by no means exhaustive enough. He may therefore be allowed to express here his obligation and thanks to all his fellow-workers; and in particular, to the Association of German Chocolate Manufacturers Dresden; its managing director, Herr Greiert; the director of the Cocoa Purchase Co, Hamburg, Herr Rittscher, who contributed the whole of the chapter headed;

Commercial Varieties of Cacao Beans further to Prof. Dr. Härtel, Chief Inspector of the Royal Research Institute, Leipsic; Dr. R. Böhme, Managing Director of Messrs. Stollwerck Bros. Chemical Laboratory, Cologne; and to Superintendent Engineer Schneider, of the firm J. M. Lehman, Dresden, among many others. Mention must also be made of the manufacturers who so kindly placed material at the editor's disposal. Let us hope that the work will meet with a success corresponding to the pains taken by the editor and publishers, and prove a really serviceable Handbook to the Chocolate Industry.

Dr. Schaeffer.

Extracts from the prefaces to the first and second editions.

The object of this work is to furnish a source of information and advice for those who are interested in the branch of industry to which it relates.

The author of this treatise has therefore endeavoured not only to describe the manufacturing processes; but he has also devoted special attention to the raw materials employed, and endeavoured to make them generally familiar by reference to the literature on the subject, as well as by providing a precise account of the chemical constituents of these substances and discussing the consequently necessary procedure to be observed in the course of manufacture. The art of chocolate making is no longer what it was a few decades ago; it has for the most part passed from small operators into the hands of large manufacturers. A short historical resumé will serve as a sketch of this development and a cursory description of some forms of apparatus which have now merely historical interest will serve to show how improvement in the industry has been effected.

Chocolate is a favourite and most important article of food, and in that sense it is subject to legal regulations for which allowances must be made, as well as for the most suitable analytical methods by means of which a manufacturer can ascertain the presence of unlawful mixtures in competing products, so that knowing the regulations in force, he may avoid any infringement of the same.

Within the ten years that have elapsed since the first edition of this work appeared, the manufacture of chocolate has undergone considerable expansion. Not only has the m o d u s o p e r a n d i been simplified and improved by the introduction of a number of new mechanical appliances, but the technique of the subject has been so extended, both from chemical and mechanical points of view, as partly to furnish a new standard in estimating and determining cacao constituents and preparations. The author has endeavoured to take due account of all these advances, and made a point of collecting the material scattered through the various professional journals, sifting or supplementing where necessary, in order that all engaged in the industry, t h e m a n u f a c t u r e r a s w e l l a s t h e f o o d a n a l y s t a n d t h e e n g i n e e r, may be in a position to derive a vivid impression of existing conditions in the chocolate manufacture, from the present volume.

In consideration of the importance which several branches of the industry have recently acquired, such as the preparation of cocoa powder, soluble cocoa, cacao butter, pralinés and chocolate creams, space has been given to descriptions of the respective details. On the other hand no attempt has been made to introduce calculations as to the cost of manufacture, since statements to that effect would possibly be rather detrimental than otherwise.

Costs of production as regards cacao preparations is subject to great variation, according to the scale on which they are carried out, so that estimates made on the basis of large operations might eventually lead to the conclusion that a small factory might be profitable, and with no better result than that of creating undue competition in prices and occasioning eventual failure. Moreover, the fluctuations in the market price of cacao and sugar are so frequent, and there is such possibility of new sources of expense, that calculations can only apply to the time when they are made; they soon become out of date, and then afford no trustworthy indication of probable profit and loss.

The section treating of legislative regulations relating to the trade in cacao preparations has undergone complete revision to adapt it to existing conditions.

To render the book more useful, an appendix has been added in which the production and composition of a few cacao preparations are treated of, providing valuable data for reference.

Dr. Paul Zipperer.

CONTENTS

First Part: **The Cacao Tree** Page

A. Tree and Beans 1

  a) Description of the Cacao Tree and its Fruit 1

  b) Geographical Distribution and History of the Cacao Tree 4

  c) Cultivation of the Cacao Tree; Diseases and Parasites 7

  d) Gathering and Fermentation 9

  e) Description of the Beans 12

  f) The Commercial Sorts of the Cacao Bean 16

    1. American Cacao Varieties 19

    2. African Cacao Varieties 28

    3. Asiatic Cacao Sorts 32

    4. Australian Cacao Sorts 33

  g) The Trade in Cacao and the Consumption of Cacao Products; Statistics 33

B. Chemical Constitution of the Bean 43

  a) The Cacao Bean Proper 43

    1. Water or Moisture 49

    2. Fat 49

    3. Cacao red or Pigment 59

    4. Theobromine 62

    5. Albumin 67

    6. Starch 70

    7. Cellulose or crude fibre 72

    8. Sugar and plant acids 73

    9. The mineral or ash constituents 73

b) The Cacao Shells   76

Second Part: **The Manufacture of Cacao Preparations**

A. **Manufacture of Chocolate**   85

  I. The Preparation of the Cacao Beans   87

    1. Storing, cleansing and sorting   87

    2. Roasting the Beans   89

    3. Crushing, hulling and cleansing   100

    4. Mixing different kinds   108

  II. Production of the Cacao Mass and Mixing with Sugar   109

    5. Fine grinding and trituration   109

    6. Mixture with sugar and spices   117

    7. Treatment of the Mixture   119

      a) Trituration   119

      b) Levigation   123

      c) Proportions for mixing cacao mass, sugar and spices   136

  III. Further Treatment of the Raw Chocolate   138

    8. Manufacture of "Chocolats Fondants"   138

    9. Heating Chambers and Closets   141

    10. Removal of Air and Division   143

  IV. Moulding of the Chocolate   149

    11. Transference to the Moulds   149

    12. The Shaking Table   156

    13. Cooling the Chocolate   162

      a) Cooling in Chambers.

      b) Cooling in Closets.

  V. Special Preparations   176

a) Chocolate Lozenges and Pastilles 176

b) "Pralinés" or coated goods 182

B. **The Manufacture of Cocoa Powder and "Soluble" Cocoa** 195

a) The various methods of disintegrating or opening up the tissues of cacao 195

b) Methods of disintegration 197

1. Preliminary Treatment of the Beans 197

2. Expression of the Fat 199

3. Pulverising and Sifting the defatted Cacao 209

c) Disintegration after Roasting 216

1. Disintegration p r i o r to Pressing 217

2. Disintegration a f t e r Pressing 224

3. Opinions to these methods 225

C. **Packing and Storing of the Finished Cacao Preparations** 228

a) General hints 228

b) Suitable storage 228

c) Machines for packing en masse 229

Third Part: **Ingredients used in the Manufacture of Chocolate**

A. **Legal enactments. Condemned ingredients** 230

B. **Ingredients allowed** 231

I. Sweet Stuffs 231

a) Sugar 231

b) Saccharin and other sweetening agents 234

II. Kinds of Starch, Flour 236

1. Potato starch or flour 236

2. Wheat starch 236

13

3. Dextrin 237

4. Rice starch 237

5. Arrowroot 237

6. Chestnut meal 238

7. Bean meal 238

8. Salep 238

III. Spices 238

    a) General Introduction 238

    b) Vanilla 241

    c) Vanillin 243

    d) Cinnamon 246

    e) Cloves 247

    f) Nutmeg and Mace 247

    g) Cardamoms 248

IV. Other Ingredients 248

    a) Ether oils 248

    b) Peru balsam and Gum benzoin 249

V. Colouring Materials 250

Fourth Part: **Examination and Analysis of Cacao
Preparations**

A. **Chemical and microscopial examination of cacao
and cacao preparations** 253

  a) Testing 253

  b) Chemical analyses 254

    1. Estimation of moisture 254

    2. Estimation of ash 255

    3. Estimation of silicic acid in the ash 256

    4. Estimation of alkalis remaining in cocoa

powders	256
5. Determination of the fatty contents	258
6. Determination of Theobromine and Caffeine	263
7. Determination of Starch	264
8. Determination of crude Fibre	266
9. Determination of Cacao husk	267
10. Determination of Sugar	269
11. Determination of Albuminates	271
12. Investigation of Milk and Cream Chocolate	272
c) Microscopical-botanical investigation	275
B. **Definitions of Cacao Preparations**	279
a) Regulations of the Association of German Chocolate Manufacturers relating to the Trade in Cacao Preparations	279
b) Final Wording of the Principles of the Free Union of German Food Chemists for the estimation of the Value of Cacao Preparations	282
c) Vienna Regulations	284
d) International Definitions	285
C. **Adulteration of Cacao Wares and their Recognition**	288
a) Introductory	288
b) The Principles	288
c) Laws and Enactments as to Trade in Cacao Preparations	291
1. Belgium	291
2. Roumania	293
3. Switzerland	294
4. Austria	298
5. Germany	301

15

Fifth Part: **Appendix**

A. **Installation of a Chocolate and Cacao Powder Factory (with 2 plates)** — 304

    1. Chocolate Factory (Table I) — 305

    2. Cacao Powder Factory (Table II) — 306

    3. Appendix containing an account of the methods of preparation and the composition of some Commercial dietetic and other cacao preparations — 306

INDEX

A. Index to literature — 319

B. Tables — 320

C. Figures — 321

D. Authors — 323

E. Alphabetical index to contents — 326

Part I.
The Cacao Tree.

A. Tree and Beans.

a) Description of the Cacao Tree and its Fruit.

The cacao tree with its clusters of red blossom and golden yellow fruits is conspicuous even in tropical vegetation. Of considerable diameter at the base, it often attains a height of eight metres. Its wood is porous and light; the bark is cinnamon coloured, the simply alternating leaves are from 30 to 40 cm. in length and from 10 to 12 cm. broad, growing on stalks about 3 cm. long. The upper surface of these leaves is bright green, and the other one of a duller colour, and slightly hairy.

The flowers, which are often covered with hairs, occur either singly or united in bunches not only on the thicker branches but also all along the trunk from the root upwards. (Fig. 1 A.)

The formation of the fruit takes place only from the flowers of the stem or thicker branches, and for a thousand flowers there is only one ripe fruit.

The flowers (fig. 1 B & C) are very small and of a reddish white colour. Calyx and corolla are five partite, the ten filaments are united at their base (fig. 1 G) and only half of them are developed to fruitful organs, such as bear pollen (fig. 1 J) in their four separate anther compartments (fig. 1 H).

The pistil is formed of five united carpels and bears in each of its five compartments eight ovules. (Fig. 1 E & F).

The fruit is at first green, and afterwards turns yellow, but with streaks and tints of red occurring; many varieties also are entirely crimson. Resembling our cucumber in size, shape and appearance (see fig. 2 A & B), it has a length of about 25 cm. and a diameter of 10 cm., and the thickness of its shell is from 15 to 20 mm. This shell is of rather softer consistency than that of the gourd, and has five deep longitudinal channels, with five others of less depth between them.

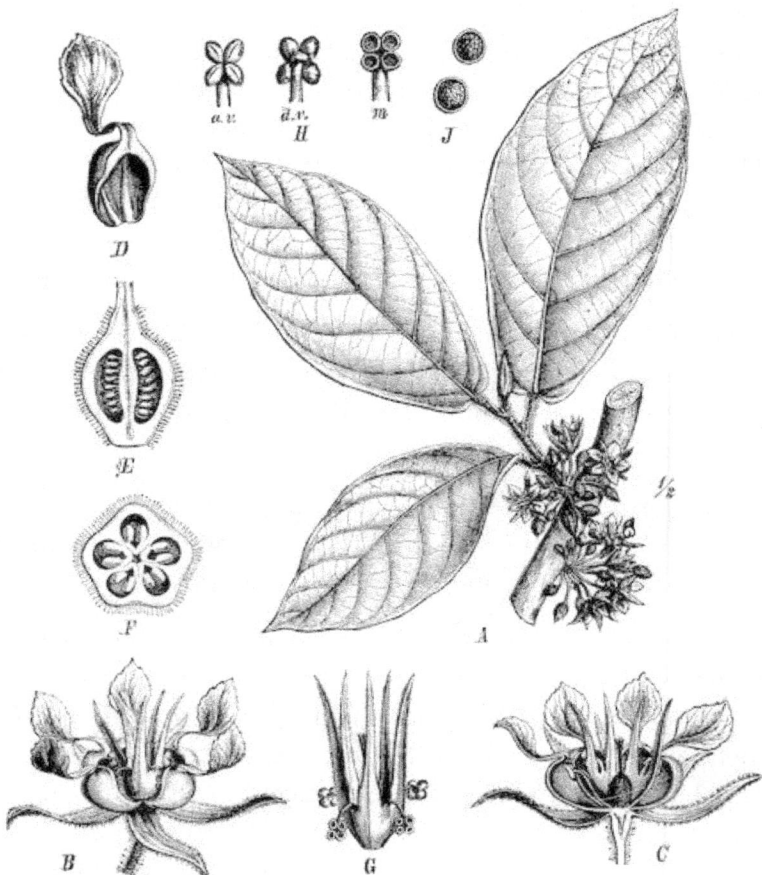

Fig. 1. (After Berg & Schmidt, Atlas.)

A Twig in bloom (1/2). B Flower (3/1). C Flower in vertical section (3/1). D Leaf of flower (6/1) E Bean-pod in vertical section (6/1) F Bean-pod in cross section (9/1). H Anther. J Pollen.

The shell encloses a soft, sweetish pulp, within which from twenty-five to forty almond shaped seeds are ranged in five longitudinal rows, close to each other. The white colour of these seeds is frequently tinged with yellow, crimson, or violet (Sec. Fig. 2 C. D. & G).

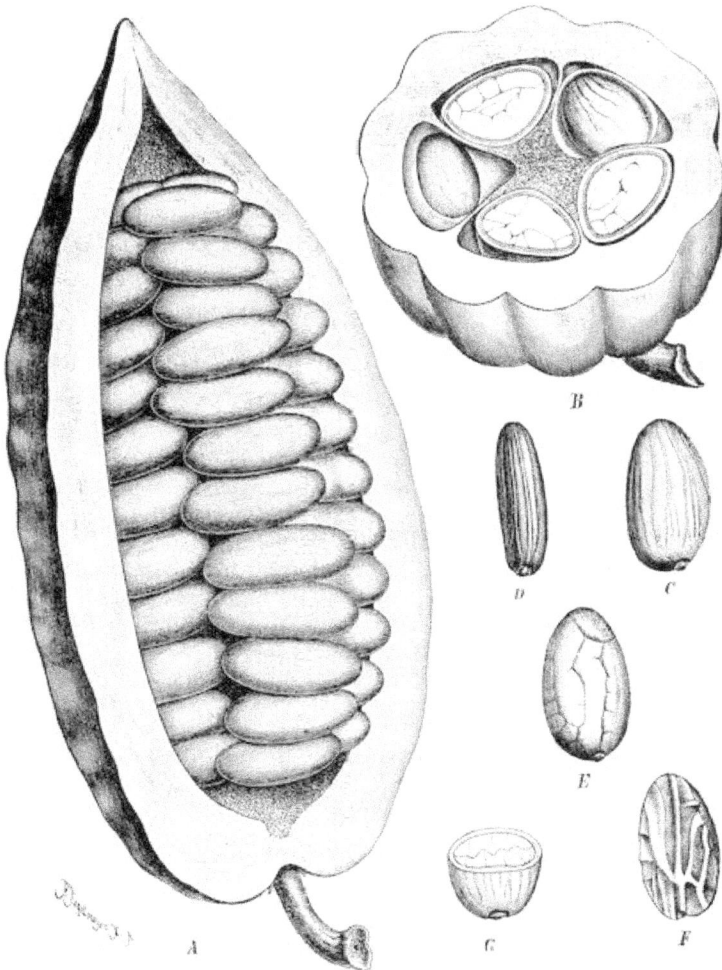

Fig. 2. (After Berg & Schmidt).

A Fruit with half of shell removed (1/2). *B* Fruit in cross section (½). *C* Side view of seed (1/3). *D* Front view of seed (1/1). *E* Seedling (1/1). *F* Kotyledon or Seed-leaf (1/1). *G* Seed in cross section (1/1).

The fruits ripen throughout the whole year, though but slowly during the dry season; and the time needed for its full development is about four months. It may be gathered at all times of the year, although there are regular gathering seasons, determined and modified by the respective climatic conditions. So, for example, we find that in Brazil the principal gathering takes place in February and July, whilst in Mexico it is in March and April. In the primeval Amazonian forests the fruit of the cacao tree is gathered and brought to market at all times of the year, wherever Indian tribes obtain.

b) Geographical Distribution and History of the Cacao Tree.

The cacao tree flourishes in a warm, moist climate. It is therefore indigenous to tropical America, from 23° north to 15° or 20° south latitude.

Consequently the area in which it grows comprises the Central American republic of Mexico down to the Isthmus of Panama; Guatemala, the Greater and Lesser Antilles, Martinique, Trinidad, St. Lucia, Granada, Cuba, Haiti, Jamaica, Puerto Rico, Guadeloupe, and San Domingo; in South America, the republics of Venezuela, Columbia, Guiana, Ecuador, Peru and the northern parts of Brazil, especially the districts lying along the middle Amazon.

In all other countries where the cacao tree now flourishes, it has been naturalised, either by colonists, or with government aid, as in Asia, where the Philippine Islands, Java, Celebes, Amboyna and Ceylon in particular are deserving of mention; and in Cameroon (Bibundi, Victoria and Buea), Bourbon, San Thomé and the Canary Islands in Africa, where the tree is sometimes found growing at an elevation of about 980 ft. above sea level. Ceylon offers an instructive illustration of the zeal with which the cultivation is carried on in some districts. According to information furnished by Mr. Ph. Freudenburg, late German Consul at Colombo, cacao had been planted in a few instances during the time Ceylon was in possession of the Dutch, but only since 1819 has seed been distributed out of the botanical gardens at Kalatura, and it was still later before planters could obtain it from those established at Peradenija. Systematic cultivation for commercial purposes was commenced in 1872 or 1873. The principal seats of cacao plantations are Dumbara, Kurunegalla, Kegalla and Polgahawella, together with North, East and West Matala, Urah and Panwila.[1] According to statistical records, the relation between the growth and export of cacao is shown by the following table, which also shows the development of its cultivation:

Year	Area under cultivation (acres)	Exports (cwts)
1878	300	10
1879	500	42
1880	3000	121
1881	5460	283
1885	12800	7247
1892	14500	17327
1895	18278	27519
1898	22500	32688
1908	39788	62186

Like all other articles of human food, cacao has a history of some interest, the most essential points of which are here summarised from the excellent work of A. Mitscherlich.[2]

A knowledge of the cacao tree was first brought to Europe

in 1519 by Fernando Cortez and his troops. He found in Mexico a very extensive cultivation of cacao, which had been carried on for several centuries. In the first letter addressed by Cortez to Charles the Fifth, he described cacao beans as being used in place of money. Cortez applied to the cacao tree the name of "Cacap", a word derived from the old Mexican designation "Cacava-quahitl". The Mexicans called the fruit "Cacavacentli", the beans "Cacahoatl" and the beverage prepared from them "Chocolatl"[3], said to be derived from the root "Cacava" and "Atl", water. This term was adopted by the Spaniards, and it gave rise in the course of time to the word "Chocolate", which is now universal.

The botanical definition of the typical form of the cacao tree, which belongs to the family BUTTNERIACEAE, is referable to Linnaeus, who gave it the name "Theobroma Cacao" (food of the gods, from "Theos", God, and "Broma", food). Probably chocolate was a favourite beverage with Linnaeus, who may have been acquainted with the work of the Paris physician Buchat, published in 1684, in which chocolate is alluded to as an invention more worthy of being called food of the gods than nectar or ambrosia. Clusius first described the cacao tree in his "Plantae exoticae". The taste for chocolate soon spread throughout Spain after the return of Cortez' expedition from the New World, not, however, without encountering some opposition, especially on the part of the clergy, who raised the question whether it were lawful to partake of chocolate on fast days, as it was known to possess nutritive properties. However, it found an advocate in Cardinal Brancatio, who described it as an article belonging, like wine, to the necessaries of life, and he therefore held that its use in moderation could not be prohibited. In 1624 Franciscus Rauch published a work at Vienna, in which he condemned the use of chocolate and suggested that the monks should be prevented from partaking of it, as a means of preventing excesses. About the commencement of the 17th century, the use of chocolate spread from Spain to Italy, where it was brought to the notice of the public by the Florentine Antonio Carletti (1606), who had lived for some time in the Antilles. The method of converting cacao beans into chocolate was also made known in Europe by Carletti, while the Spaniards had kept it a secret. Under Theresa of Austria, wife of Louis XIV, the habit of taking chocolate appears to have become very common in France after the partial introduction of cacao by importation from Spain. The first cacao imported from the French colony of Martinique arrived in Brest in 1679 in "Le Triomphant", the flagship of admiral d'Estrées. Opinion in France as to chocolate was then divided: Madame Sévigné, once an admirer of chocolate, afterwards wrote to her daughter: "il vous flatte pour un temps et puis il vous allume tout d'un coup fièvre continue qui vous conduit à la mort", a theory which nowadays must necessarily be regarded as ridiculous.

Chocolate was in general use in England about the middle of the 17th century. Chocolate houses, similar to the coffee houses of Germany, were opened in London. Bontekoë, physician to the Elector Wilhelm of Brandenburg, published in 1679 a work entitled "Tractat van Kruyd, Thee, Coffe, Chocolate," in which he spoke very strongly in favour of chocolate and contributed very sensibly to the increase of its consumption in Germany. The first chocolate factory in Germany is said to have been erected by Prince Wilhelm von der Lippe about the year 1756 at Steinhude. This prince brought over Portuguese specially versed in the art of chocolate making.

c) Cultivation of the Cacao Tree; Diseases and Parasites.

The first information regarding the cultivation of the cacao tree in Mexico is that obtained on the invasion of the country by the Spaniards. Prior to that time there is a total absence of anything definite. The tree flourishes best in situations where the mean temperature is between 24° and 28° C. The farther the place of cultivation from the equator the poorer is the product. The other most essential conditions are long continued moisture of the soil and a soft, loose texture with abundance of humus, and above all, shelter from the direct rays of the sun. For these reasons, planters select for their cacao areas ground the virgin soil of which has not been exhausted by the cultivation of other plants. The plants are either raised in a nursery until they reach the most suitable age for transplanting, or the seeds are sown on the ground selected for the plantation. The transport of live seed for new plantations is attended with some difficulty, since the seeds very quickly lose their vitality. C. Chalot[4] recommends that this vitality be preserved by gathering the fruit before it is perfectly ripe, immersing it in melted paraffin oil, and then wrapping it in paper; on which the fruit may be transported without losing any of its nutritive qualities.

In the sheltered valleys of tropical countries, where the soft soil, rich in humus, is kept constantly moist by large rivers, the cacao tree blossoms throughout the whole year. When growing wild it is generally isolated under the shadow of larger trees; when cultivated, the young plant is placed under the shelter of banana trees, and at a later period of its growth shelter is provided by the coral (called Erythrina corallodendron or Erythrina indica), further known as "Coffie-mama" among the Surinam Dutch and madre del cacao among the Spaniards. Yet this tree, like the Maniok, is said not to enjoy so long a life as the cacao plant, which sometimes reaches an age of forty years. On this account the Castilloa or also Caesalpina dashyracis have recently been recommended as a more lasting protection. The fact that it does not lose its leaves during the dry season (e. g. on Java, during the East Monsoon) is an additional advantage.

A cacao plantation requires a considerable area, in the proportion of 50 hectares for 20,000 trees. The quantity of fruit to be obtained from that number of trees, as an annual crop, would be worth from £1,200-1,300. In planting the seeds, they are set in rows that are from 8 to 10 m. apart, four or five seeds being planted within from 1 to 2 m, the shading trees being planted between the rows. Of each five seeds planted the greater number often fail to germinate, either in consequence of unfavourable weather or as the result of attacks by insects etc.; but if more than one plant grows, the weaker ones are pulled up. Until the plants are two or three years old, they are protected by a shed open at one side, and they are transplanted after they have attained a height of 3 ft. The chief enemies of tropical cultivation — weeds, aerial roots, insects, bacterial infection — have to be provided against continually, so as to prevent damage; accordingly if the ground be not moist enough, it should be systematically watered, and so drained if marshy, for the tree requires most careful nursing if it is to develop into a prolific fruit-bearing specimen. The seed germinates about fourteen days after being planted; but flowers are not produced till after 3 or 5 years. After the tree has once born fruit, which may occur at the end of the fourth year it often continues to do so for fifty years. The tree is most prolific when from twelve to thirty years old.

As in the case of all cultivated plants and domestic animals, the existence of which does not depend on the principal of natural selection, and among which life is not a continuous development of endurance in the face of adverse

25

elements, the cacao tree has its peculiar diseases. Indeed, it would seem as though it were beset by all vermin extant. The reader may obtain some idea of the extent of the damage done to cacao plantations by such noxious agents, if he turns up the clear and exhaustive account published by the Imperial Biological Institute for Agriculture and Forestry (Germany).[5] Unfortunately we have not space here to mention more than a few of the most frequently occurring and important diseases, such as the GUM DISEASE, which is especially destructive, gum formations in the wood tissue and bark of the tree eventually killing it. Next to be dreaded are the various fungus growths, cancers and cancer-like incrustations ("Krulloten") and broom formations. It often happens that specii of beetle attack the tree, causing decay and rot to set in; such e. g. are the wood-borer, bark bug, and woodbeetle. Other parasites, again, do not destroy the whole tree, but are equally detrimental, as they also preclude all prospects of a harvest. Fruit rot and its like, fruit cancer, and cacao moths, are notorious in this connection. There are also several larger creatures which betray a preference for the nutritious fruit of the cacao tree, various species of rat, and the squirrel, which unite to make the planter's life a burden.

d) Gathering and Fermentation.

The gathering of the fruit is effected by means of long rods, at the end of which is a semi-circular knife for cutting through the stalk. The fruits are then split in two, the beans separated from the surrounding pulp and spread out on screens to dry, or exposed to the sun on bamboo floors. Beans so prepared are described as unfermented.

In most lands where cacao is cultivated, another process is adopted, calculated to heighten the flavour of the fruit and develop its nutritious constituents. The newly gathered beans are first partially freed from the fruity substances always adhering, then piled up into heaps and covered with banana skins or cocoa-nut matting, in order that they may be shut off as far as possible from all atmospheric influence, and so left for some time, while the chemical processes of warming and fermentation are gradually consummating. This procedure is alternated with repeated exposures to the sun, according to the maturity and species of the cacao bean, and the prevailing weather conditions; though details as to the length of time and number of repetitions necessary to the production of a marketable article still await determination.[6] It may be taken as a general rule that fermentation should proceed till the bean, or rather the cotyledon, has acquired the light brown colour characteristic of chocolate. This principle is nevertheless often violated, especially as loss of weight in the bean is often intimately connected with complete fermentation. Unsufficiently fermented varieties, but which were fully ripe when gathered, develop a violet colour during this process; it is possible for them to pass through what is known as "After fermentation" before reaching the factory. This is not so in the case of beans developing from unripe fruit, for obviously the valuable constituents of the cotyledon are here not prominent, and scarcely calculated to ferment properly. Such can be recognised by their betraying a bluish grey colour in the drying processes, and the soft and smooth structure which they then acquire. A normal progress of fermentation is indicated where the interior of the mass of beans registers, on the first morning after gathering, a temperature not exceeding 30-33° C, 35-38° on the second day, and on the third morning a temperature not exceeding 43° C. If the outer shells are marked, the heating has been too severe. In countries where the harvest season suffers from the periodical rains, drying over wooden fires[7] is often resorted to. The value of many specimens is hereby greatly diminished when the roasting is carelessly managed, for the smoke must on no account be allowed to come into contact with the bean. Yet "Smoky" lots among the St. Thomas, Accra, and Kameroon sorts were formerly much more frequent in commerce than now, for the planter has learned to avoid this evil. After they have been fermented, the beans are washed, or trodden with the naked foot, in some countries, and so cleansed from the pulp remains still adhering. They are then allowed to dry in the open air, and packed into sacks; contact with metal or stone is strictly to be avoided, which as good conductors of heat and rapid cooling agents are most disadvantageous. Instead of piling the beans up in loose heaps, they may be fermented in "Tanks" made of wood, and where possible, provided with partitions. According to Kindt, cedar wood has been proved best for this purpose, because of its enormous resisting capacity. It used to be thought that in fermentation ensued a germination of the seed,[8] as in the preparation of malt; but this idea has been proved erroneous. The contrary is rather the case, for the process almost kills the seed; and when the sensitiveness of the latter is taken into consideration, and also the fact that it only develops under the most favourable conditions, it must be allowed that the statement contains an obvious truth. Yet chemical change does take place in the fermentation of the seed; but as to its precise nature, owing to the lack of scientific research on the scene of operations, we are still unable to dogmatise. It would therefore be useless to discuss the manifold theories and speculations bearing on this point, and waste of time to discuss the various kinds of fermentation and the chemical processes therein involved. Yet it may almost be taken for granted, that the fresh-plucked bean contains a so-called glycoside[9] which decomposes into grape sugar, into an equally amyloids colour stuff (the so-called cacao-red), and the nitrogeneous alkaloids Theobromine and Kaffein; a change probably incidental to the fermentation.[10] The sugar might further split up into Alcohol and Carbonic Acid Gas, although this is by no means established.

Whilst we have lost our bearings as far as the chemical aspect of this process is concerned, we are much more firm in respect to the biological, thanks to researches which Dr. v. Preyers has conducted on the spot in Ceylon. Preyer's[11] experiments leave absolutely no room for objection, and it can safely be accepted that there are no bacteria present in fermentation, but a fungus-like growth rich in life, a kind of yeast by him called Saccharomyces Theobromae, and described in passing;[12] facts which constitute the gist of his findings. He further establishes that the presence of bacteria often noticed is absolutely undesirable, and that better results are obtained when all life is energetically combated, and especially these bacteria. We should, then, be confronted with the same phenomenon in the preparation of cacao as are already met with in beer brewing, and the pressing of wine and which are still waited for in the preparing of tea and tobacco.

The kernel of the fresh bean, "Nips", is white and has a bitter taste and alternates in colour between whitish yellow, rose and violet; the mere influence of solar heat is sufficient to produce the brown cacao pigment, but drying is not so effective as fermentation in removing the harsh bitter taste and hence fermented beans are always to be preferred. These have often acquired a darker colour in the process, their weight is considerably diminished, and their flavour modified to an oily sweetness, without losing an atom of the original aroma[13].

Commercially and for manufacturing purposes only the seeds of the cacao tree are of importance. The root bark is said by Herr Loyer of Manila to be of medicinal value as a remedy for certain common female complaints and is employed by the natives of the Philippine Islands as an abortifacient. According to Peckoldt[14] the fruit shell contains a considerable amount of material that yields mucilage and might therefore be utilised as a substitute for linseed.

e) Description of the Beans.

The varieties of the cacao tree which yield the beans at the present time occurring in commerce are.

Theobroma cacao, Linné the t r u e c a c a o, spread over the widest area, and almost e x c l u s i v e l y c u l t i v a t e d o n p l a n t a t i o n s with many varieties (Crillo, Forastero etc.) and Theobroma b i c o l o r, a party-coloured cacao tree the seeds of which are mixed with Brazilian and Caracas beans.

Theobroma speciosum Wildenow, which yields, like Theobroma cacao, Brazilian beans (magnificent tree).

Theobroma quayanense, yielding Guiana beans.

Theobroma silvestre or forest cacao.

Theobroma subincanum, w h i t e - l e a v e d - c a c a o, and Theobroma microcarpum, s m a l l - f r u i t e d c a c a o a r e m e t w i t h a s a d m i x t u r e s in Brazilian beans.

Theobroma glaucum, g r e y c a c a o fruits of which variety are found among Caracas beans.

Theobroma angustifolium the n a r r o w - l e a v e d and Theobroma ovatifolium, o v a l l e a f may be regarded as characteristic of Mexican cacao.

Before describing the commercial kinds of cacao, a knowledge of which is of first importance to manufacturers, it is desirable to consider the beans in regard to external form and microscopial structure, in order that the use of some indispensable scientific expressions in the subsequent description of particular commercial kinds of cacao may be intelligible.

The bean, page 3 Fig. 2 C-G, consists, according to Hanousek[15], of a seed-shell, a seed-skin and the embryo or kernel with the radicle. The oval-shaped seed is generally from 16 to 28 mm. long, 10 to 15 mm. broad and from 4 to 7 mm. thick. At the lower end of the bean there is a depressed, flattened and frequently circular hilum visible, from which a moderately marked line extends up to the apex of the bean where it forms the centre of radiating longitudinal ribs — vascular bundles-extending to the middle of the bean through the outer seed-coating back to the hilum.

The outer seed shell (cf. Fig. 3) is of the thickness of paper, brittle, scaly externally and reddish brown, lined with a colourless translucent membrane peeling to the so-called silver membrane (previously but falsely known as seed envelope) and penetrating into the convolutions of the kernel in irregularly divided folds. The shells of some of the better sorts of beans, such as Caracas, are frequently covered with a firmly adherent, dense, reddish-brown powder, consisting of ferruginous loam originating from the soil on which the beans have been dried and serving as a protection against the attacks of insects. But opinions are divided as to, the utility of this process.

The fermented kernel consists of two large cotyledons occupying the whole bean; it is of fatty lustre, reddish grey or brown colour and often present a superficial violet tinge; and under gentle pressure readily breaks up into numerous angular fragments the surfaces of which are generally bordered by the silver membrane. The fragments can be easily recognised when laid in water. At the contact of the lobes there is an angular middle rib and two lateral ribs are connected with the radicle at the broader end of the bean. The ripe fresh-gathered cacao-kernel is undoubtedly white and the reddish brown or violet pigment is formed during the fermenting of the bean. But there is also a white cacao, though seldom met with. According to information furnished by Dr. C. Rimper of Ecuador, it is of rare occurrence and is not cultivated to any great extent. In Trinidad also a perfectly white seeded cacao, producing large fruit and fine kernels, was introduced from Central America by the curator of the Botanic Gardens in 1893.

The microscopic structure of the shell, Fig. III., presents no remarkable peculiarity that requires to be noticed here.

The delicate inner membrane (fig. 3) coating the cotyledons and penetrating into their folds consists of several layers. Connected with it are club-shaped glandular structures, fig. 4, consisting of several dark coloured cells that are known as the Mitscherlich particles According to A. F. W. Schimper[16] they are hairs fallen from the epidermis (fig. 4) of the cotyledon and do not originate, as was formerly supposed, in the inner silver membrane.

These structures, named after their discoverer, were formerly supposed to be algae, or cells of the embryo sac, unconnected with the tissues of the seed cells. They are, however, as true epidermoid structures, similar to the hairs of other plants.

32

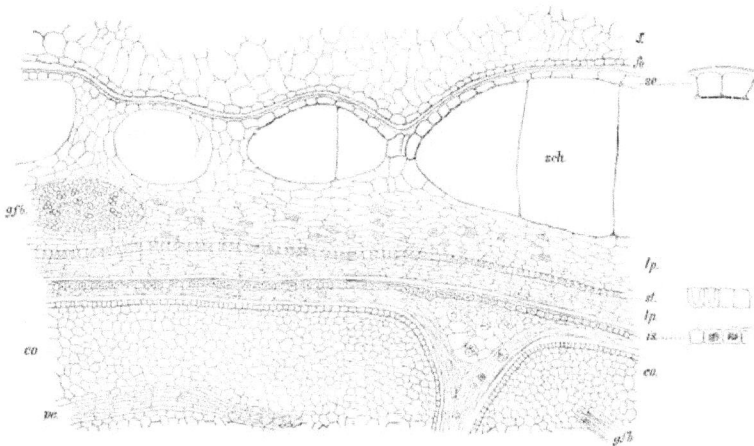

Fig. 3. Cross Section of Shell of Cacao Bean (Tschirsch).

gfb vascular bundles	*fe* endocarp, or inner coat of fruit	*st* sklerogenous, or dry cells
co cotyledon	*se* epicarp, or skin	*is* silver membrane
pc ducts	*sch* mucilagenous, or slime cells	*co* cotyledon
f pulp	*lp* parenchyma, or cellular tissue	*gfb* vascular bundles

These Mitscherlich particles are not only characteristic of the seed membrane but also of the entire seed as well as the preparations made from it. Wherever cacao is mixed with other materials, its presence may be ascertained by microscopical detection of these structures, which are peculiar to cacao.

In the large elongated, hexagonal cells of the seed membrane there are two other structures to be seen with the aid of high power (250 fold), one appearing as large crystalline druses, while the other consists of extremely fine needles united in bundles.

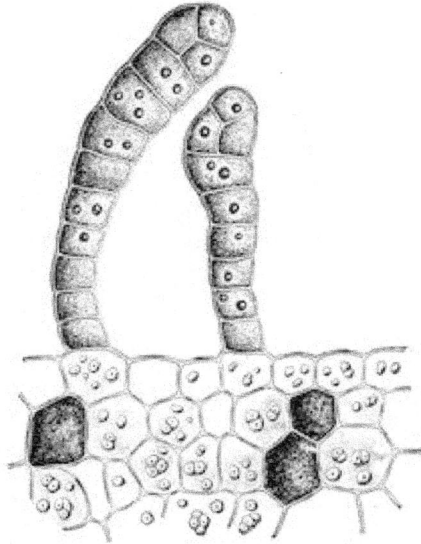

Fig. 4.
Cross section of the cotyledon, showing "Mitscherlich particles" (Moeller).

By addition of petroleum spirit the former, consisting of fat acid crystals, are dissolved, the latter, remaining unaltered, are considered by Mitscherlich to be theobromine crystals, since their crystalline form closely resembles that of theobromine. A more scientific explanation has not been forthcoming.

The cotyledons are seen under the microscope to consist of a tissue of thin walled cells, without cavities, lying close together, and here and there distributed through the tissue, cells with brownish yellow, reddish brown, or violet coloured contents. These latter are the pigment cells which contain the substance known as cacao-red and analogous to tannin; it, together with theobromine, gives rise to the delicate taste and aroma of cacao. The other cells of the tissue are filled with extremely small starch granules the size

34

of which rarely exceeds 0.005 mm.; with them are associated fat, in the form of spear-shaped crystals, and albuminoid substances.

In order to discriminate between these substances they must be stained by various reagents. According to Molisch[17], theobromine may be recognised, in sections of the seed, by adding a drop of hydrochloric acid and after some time an equal drop of auric chloride solution (3 %) After some of the liquid has evaporated, bunches of long yellow crystals of theobromine aurochloride make their appearance. On addition of osmic acid the fat is coloured greyish brown. On addition to the microscopic section a drop of iodine solution, or better iodozine chloride, the starch becomes blue, while albuminous substances are coloured yellow. Cacao starch granules are very small and cannot well be mistaken for other kinds, except the starch of some spices such as pimento or that of Guarana, prepared from the seeds of Paulinna sorbilis. According to Möller the blue iodine colouration of cacao starch takes place very slowly and it is probably retarded by the large amount of fat present; but the point has been contested by Zipperer and later investigators.

In order to make the starch granules of cacao and the cells containing cacao-red distinctly visible under the microscope, it is advisable to immerse the section in a drop of almond oil, because the addition of water renders the object indistinct in consequence of the large amount of fat present. Another excellent medium for the microscopic observation of cacao is the solution of 8 parts of chloral hydrate in 5 parts of water, as recommended by Schimper.[18]

By these means it may easily be seen that the pigment or cacao red in different sorts of cacao varies more or less in colour.

To complete the account of the microscopic characters of the cacao cotyledon, mention must be made of the small v a s c u l a r   b u n d l e s, generally spiral, that are distributed throughout the tissues of the cotyledons and are readily made visible by adding a drop of oil or a drop of chloral hydrate solution.

f) The Commercial Sorts of the Cacao Bean.

Mindful of Goethe's dictum: F r i e n d ,   t h e   p a t h s   o f t h e o r y   a r e   u n c e r t a i n ,   a n d   h i d   i n   g l o o m, we propose to devote this chapter to an exclusively practical discussion of the commercial value of raw cacao, and from the merchant's point of view.

Such differences of opinion prevail in manufacturing circles as to the possible uses of each separate sort, that for this reason alone any other than a purely geographical classification would scarcely be feasible. But apart from this, varying as it does with the protective duties imposed, the commercial value of cacao can by no means remain a universal constant; and it must be noted that variations in the national taste serve to heighten its instability.

This latter circumstance also causes a deviation from the nearly related principal that the Motherland becomes chief consumer of the varieties grown in her colonies. The cacao sorts of the English Gold-Coast running under the collective name of A c c r a, have taken complete possession of the German market; Trinidad cacao enjoys immense popularity in France, and the Dutch pass on the larger part of their Java importations to other consuming nations. As regards this latter sort, however, the fact they are chiefly employed as colouring and covering stuffs for other cacaos must be taken into consideration.

In most cases either the producing country or a principal shipping port gives its name to the different sorts. Yet paradoxical exceptions will at once occur to the reader. The inferior and mediocre Venezuelan varieties of the Barlovento district shipped from La Guayra are generally denominated as C a r a c a s, notwithstanding the fact that the capital of the republic Venezuela, situated as it is 1000 metres above sea level (being about 3300 feet), and therefore quite outside the cacao zone, has practically no connection with the cacao trade. The collective name, Samana still holds good for the cacaos of the Dominican republic, at least in Germany, although this outlet of a tiny mountainous peninsular has long ceased to export any but very insignificant quantities. Consequently, and rightly, the French merchant specifies these sorts as S a n c h e z, adopting the name of the principal cacao exporting port of the republic. Arriba, the choicest product of Ecuador (port, Guayaquil) takes its name from the Spanish word arriba, above, the plantations being situated along the upper sources of the Rio Guayas (to wit, the rivers Daule, Vinces, and Zapotal). Other Guayaquil cacaos are named after the rivers (Balao, Naranjal) and districts (e. g. Machala) where they are most cultivated.

As in the case of so many other cultivated plants, distinguishing characteristics of the various sorts are not only determined by the different species of tree, but are rather and principally dependant on the combined effect of physical and climatic conditions. So whether the seedling Criollo, the splendid Creole bean native to Venezuela, belongs also to the more fruitful Forastero species (spanish forastero, foreign), a variety less sensitive and consequently commoner, is a problem which can only claim secondary consideration.

Apart from the geographical influences mentioned, method and nicety of procedure are of prime importance in the preparation of the cacao sorts. Yet technically perfect implements do not always prove the best means to an attainment of this end; it being a fact recorded by experience that the chemical constituents of the cacao bean reach their fullest developement in such simple and primitive processes as, e. g. are still patronised in Ecuador and Venezuela. It is scarcely necessary to observe that these simple and primitive methods postulate nicety and carefulness, which failing, there will be no lack of defects in the cacao prepared. On the Haiti/Domingo island, e. g. a variety of cacao is harvested which is in itself very profitable, as stray specimens finding their way to the market testify, but which as an article of commerce proves most unreliable, being generally brought on the market in such an unprepared state, that fermentation first takes place on the sea voyage, and then of course only in insufficient measure. During this period appear those disagreeable and accompanying symptoms technically known as "Vice propre" and the beans, which were not completely ripe in the first place, do not develop further, and greenish breakings in the skin become pronounced, and remain a source of terror to the manufacturing world. All attempts made in European interests to bring about an alteration in this deplorable state of affairs have hitherto been lost on the indolence of the native planters. Indeed, until the political and economical conditions prevalent among the mixed Negro population of Haiti/Domingo are thoroughly reformed, no perceptible improvement can be expected in the qualities of the Samana and Haiti cacaos, for which reason, with rapidly disappearing exceptions, there are scarcely any well organised plantations in these parts.

Turning to the Old World, we find in the West African Gold Coast a typical example of the possibilities of cultivation on a small scale, under proper and competent guidance, and with primitive processes; for not only as far as quantitative progress is concerned, but also in respect to quality, the varieties produced by the natives of this English colony improve from year to year. Kameroon, a district which like the Gold Coast has only taken to the cultivation of cacao of late years, provides us with an exactly opposite instance. Here the plantation system has been in force right from the commencement of the industry, with all its technically perfected implements, yet nevertheless the perfecting of the cacao proceeds very slowly, and it will be a long time before the produce of this land can lay any serious claim to specification as a variety for consumption. Its large proportion of acid ingredient has been above all detrimental, almost completely precluding its use as any other but a mixing sort, although some plantations have been yielding comparatively mild cacaos now for several years. We cannot stay to discuss the problem of causes in this instance, and whether the fact that the Forastero species has been exclusively planted prejudices the developement of the cacao, or the climatic conditions, must remain an open question. Let it be noted in passing that the Forastero Bean has taken universal possession of Africa, as well in Kameroon, as in the Gold Coast, on the island of St. Thomas and also in the Congo Free State. The Bahia cacao, again, owes its origin to the Forastero seedling.

We will refrain from any further elaboration of this introduction, however, so as not to anticipate the following review of the various commercial sorts of cacao.

f) I. American Cacao Varieties.

A. Central America.

We begin with

Mexico, the classical cacao land, scarcely of importance to the general trade, as the greater part of its entire produce, comprising about three thousand tons yearly, is consumed in its native country. Of the other Central American states, next to

Nicaragua, whose large Venezuelan-like beans find their way to the Hamburg market from time to time,

Costa Rica is above all worthy of mention. This state began to export its home produce in 1912, averaging for that year about 60 tons; and in 1909, the export had already increased to 350 tons, mostly to England and North America, through the shipping port called Port Limon.

B. South America.

Columbia. From this republic come two distinct sorts; the rare, rounded, and native

Cauca bean, which is nearly related to the Maracaibo variety, and which cultivated along the Magdalena river is in the main shipped from Baranquille, on the Caribbean sea, occasionally also from Bueneventura on the Pacific coast; and then the

Tumaco Cacao so named from the small shipping port on the Ecuador border, which resembles the inferior sorts of the Ecuador coast.

Cauca-and Tumaco-cacaos are only seldom free from defective beans and worm-eatings, probably less caused by the primitive processes of preparation than the difficult means of communication in this country. Then also considerable quantities are retained for home consumption.

Ecuador is the home of the cacao richest in aroma, the country which first developed the plantation system on a

40

large and well organised scale, and which was still at the head of cacao-harvesting lands a few years ago, with a yearly produce of about 32,000 tons. Yet although it had increased this amount to 40,000 tons in the year 1911, Ecuador can only take second rank among cultivating lands, the Gold Coast coming first. The following and most valuable varieties are embraced under the name of the chief shipping port.

GUAYAQUIL. They are:

1. A r r i b a, i.e. above, these cacaos coming from the upper tributaries of the rio Guaya (the rivers Daule, Vinces, Publoviejo, and Zapatol). The Arribas, like the Guayaquil cacaos generally, are chiefly used in the preparation of cacao powders. They form e. g. the principal constituents of the Dutch cacao powders, especially the so-called superior Summer-Arriba, harvested from the month of April to July. All that is gathered in other seasons falls into the general class "Arriba superior de la época

The cacaos of the months immediately following on Summer, the r e b u s c o s, after crop, are as a rule the most inferior varieties of arriba, whilst the Christmas harvest of the months of January and February (cosecha de Navidad) often yields quite excellent sorts.

2. M a c h á l a, second in importance among the Guayaquil sorts, rather more fatty than the ariba, and differing from this again in Aroma and the colour of its kernel, which is of a rather darker brown. Chief cultivation occurs in the low lying land bordering on Peru and lying opposite the island of Jambeli, where the prevailing climatic conditions are quite different from those in the arriba districts, although these are not far removed. August and September are the harvest months for Machala. Ten years ago this sort was shipped in large measure from the then newly created harbour Puerto

41

Bolivar. But since large ocean going steamers no longer call there, it now takes the more roundabout route via Guayaquil.

3. B a l á o. This variety can be described as a mean between Machala and Arriba. It has some of the characteristics of both, the bean being somewhat rounder.

4. N a r a n j a l and T e n g u é l are likewise subdivisions of the foregoing, except that the bean is here much larger and flatter. As the production of all three sorts, and especially of Balao, is substantially greater than what finds its way to the market, we may reasonably assume that a large proportion is used for mixing purposes, and sails on commercial seas, as it were, under false colours. Cultivating district: the Machala district situated along the Jambeli canal, and the stretch of coast watered by the rivers Balao and Naranjal.

5. Pegados (i. e. stuck together) or Pelatos (balls) is the description of the cacaos comprised of series of 4-10 beans rolled together, generally developing from overripe fruit. They experience a particular kind of fermentation, apparently the result of the fruity substances still evident, which gives the light coloured kernels a soft aromatic flavour. For several years these sorts have rarely been seen on the European market, they being generally reserved for home consumption.

6. O s c u r o s, i. e. dark coloured, a refuse sort rightly viewed with suspicion in manufacturing circles—Pelotas soaked in water, or beans left in the clefts and fissures of the drying chamber floors.—The black shell of the bean encloses a brownish and dirty-looking kernel, the colour sometimes approaching black: the whole bean giving a disagreeable impression, as it is often disfigured with mould, and possessed of a disagreeable odour. For several years this variety served the "crooks" of the commercial world as

mixing material for the so-called "flavouring" of Machala, but it now again appears as a distinct sort.

The shipping port for all these cacao sorts is Guayaquil; though other harbours also handle valuable varieties. Such, for example, are

a) B a h i a   d e   C a r a q u é z, and the small haven of Manta lying south of this town, which deals in a sort resembling a blended Machala-Balao, though occasionally light brown in appearance and of aromatic flavour. This cacao is generally labelled as C a r a q u é z for short, and is to be distinguished from C a r a q u e, the French term for Caracas cacao.

The chief harvesting months are June and July; the April-May arrivals, however, are usually better, as the setting-in of the rainy season increases the difficulties of drying. The harvest in 1909 reached 3,000 tons, and is normally from 2000 to 5,000 tons yearly.

b) E s m e r a l d a s, similar to the foregoing, but of perceptibly inferior output, possesses only a very insignificant yield (about 150 tons a year), and this in spite of the cultivating capacities of the interior.

P e r u, the most southerly producing land on the west coast can likewise only boast of a very insignificant yield, chiefly destined for home consumption.

B r a z i l, with its two great sorts for consumption, Bahia and Para cacao, and a yearly production of round 33,000 tons, has from the years 1906-1909 far outrun all other harvesting lands. Yet although it was able to increase this to 36,250 tons in 1911 it must nevertheless take second place among cultivating lands, the Gold Coast and Ecuador preceding.

A most important factor on the market is included under the specification B a h i a-cacao. Here again the shipping port has given its name to the cacao sort. It is harvested in three southerly situated districts, Ilheos, Belmonte, and Canavieiras, and is despatched to Bahia from harbours of the same name, in sailing vessel which sometimes ship a thousand sacks.

Ilheos despatches the inferior of the two principal varieties "Fair fermented" and "Superior fermented" that is, the first-named, and so furnishes two-thirds of the Bahia crop. The cacao areas in the district of Ilheos are situated on rather high and mountainous ground, where arresting atmospheric conditions often predominate. Also the absence of any waterway whatever renders it a necessity to despatch the cacao to Bahia on beasts of burden, which during the rainy season can scarcely find a footing on the beaten tracks. It is, then, the unfavourable atmospheric conditions, combined with a certain carelessness on the part of the planter in the preparing processes, which prejudices the otherwise excellent quality of the Bahia bean, and more especially in the months of June, July and August.

At this period it is no rarity to find from 10 to 20 percent of waste beans, and in general only the December-February months offer anything approaching a guarantee as to quality. But here no hard and fast rule can be adduced.

Belmonte and Canavieiras are the districts of the "Superior fermented" cacaos. The lower lay of the land is responsible for other climatic conditions, and in addition, both harbours here are situated at the mouths of rivers which afford an easy and sure means of transport. So the cacao, which is also better roasted,—a few planters even drying in ovens—reaches the market in a much better condition, and fetches at least from 3-4 sh. a cwt. more than the "Fair Fermented" variety.

In all three districts, the beans are prepared in wooden boxes, covered with banana skin, in which the Ilheos variety is allowed to ferment from 2-3 days, and the superior from 2-5 days: this after they have been well shaken up. In Belmonte considerable drying takes place on the sand there deposited by the river in large quantities.

The harvesting is generally reckoned from April 1st. to March 31st. In June and July is the intermediate harvest, whilst the months from October to February supply the bulkiest crops.

The Bahia district yields yearly about 33,500 tons, a fourth part of which is devoted to the consumption of the United States, the remainder chiefly going to Germany, France and Switzerland. The return is still on the increase, and large stretches of land await cultivation.

P a r a cacao is the denomination of all those sorts shipped from the tracts of land lying along the banks of the Amazon and its mighty tributaries, more especially from Manaos and Itacoatiara, through Para, a port situated on the eastern arm of the delta. These varieties may be classed as intermediary between Bahia and good Sumana. The yearly yield (harvest months June-August) amounts to about 5,000 tons, a comparatively small figure in view of the enormous expanses capable of planting, where the cacao tree at present grows wild, or at least uncultivated. It is true that the returns for 1891 reached 6,500; only to be diminished by half in 1908. France is by far the chief country consuming Para cacao; the sort not meeting with especial favour in other states.

G u i a n a . Of the three colonies belonging to France, Holland, and Great Britain respectively, which go under this name, only the intermediate one, Dutch Guiana, is of importance in the world's cacao trade. It comes into consideration under the name of

S u r i n a m cacao. The yield, which should in normal years amount to about 3,000 tons (1899 providing the record with approximately 4,000 tons), has been considerably impaired by tree diseases and parasites. The return for 1904 only amounted to 850 tons, for example. But meanwhile Holland had hit upon excellent measures to battle against the enemies of the tree, and the years 1909 and 1910 had in consequence already improved this to 2,000 tons. The bean has some resemblance to the Trinidad bean, as far as quality is concerned.

V e n e z u e l a, one of the earliest cultivating lands, is the home of the Criollo bean, and of the most splendid specimens of bean in general, sorts which play a prominent part in the Chocolate Manufacture. The Venezuelan bean is rather long and round, and its kernel of a beautiful light brown, with a mild sweet flavour. Unfortunately the plantations have recently been interspersed with Forastero or Trinidad-Criollo trees—called in Venezuela "Trinitarios because brought over from Trinidad, a species which requires less attention and bears more fruit, but which just on that account supplies commoner and mediocre beans, slowly fermenting, and often developing a violet hue. The preparation is here of the simplest; the beans e. g. are dried on clay-covered floors, and in rainy weather earthy fragments often adhere to them. Yet such "Patios" or "Then-dales", (clay floors) are only in use on the small "haciendas" (plantations). The colouring of the Venezuelan bean with an ocre-like earth constitutes an especial peculiarity. It is adopted in particular for the medium and finer sorts. The earth is mostly sent from the neighbourhood of Choroni to the two large shipping ports Puerto Cabello and La Guayra, where the colouring or "Earthification" of the cacaos to be exported ensues. The earth, varying in colour from a dirty yellow to brick-red, is mixed to a thin paste with sea-water, and afterwards placed in the sun on large sieves, or spread over cement floors. Where the colouring takes place immediately on the plantation, the yellowish brown earth everywhere available is utilised; and where sea-water cannot be obtained, as on the Rio Tuy, for example, there the beans are coloured with a mixture prepared from crushed and almost liquid cacao fruits and this same yellowish brown earth, as the use of fresh water is thought to afford but inferior protection against mould growths. Such juice-coloured cacaos, and occasionally also the Ocumare sorts, are often covered with a rather thick earthy crust. Professional opinion concerning the utility of this colouring varies greatly. In France, the principal country consuming Venezuelan cacao, it is still maintained that the thin earthy crust not only enables the bean to resist the penetration of mildew, but also admits of a kind of after-fermentation, together with developement and preservation of the most valuable constituents of the cacao bean. Colouring is then the rule for the finer Caracas sorts, and all varieties shipped through Puerto Cabello; it is also in use at Carupano, for export to Spain.

The Venezuelan cacaos are divided as follows, and with one exception take their names from the chief shipping ports, to which they are brought in small sailing vessels tapping the villages dotted along the coast.

1. M a r a c a i b o cacao, the noble, large, and always uncoloured bean found on the shore of Sea of Maracaibo.

2. P u e r t o C a b e l l o quite the finest of all cacao sorts, with the following sub-classes, each named

after tiny harbours in the vicinity: Chuáo, Borburato, Chichiriviche, San Felipe (coloured with its own peculiar light brown earth) Ocumare, Choroni.

3. C a r a c a s cacao, exceptionally so-called, although quite a small proportion, namely that brought over the mountains from the Rio Tuy district in donkey caravans, now touches the republican capital. La Guayra, rather, is the shipping port for the so-called Caracas sorts, to which belong all the cacaos from the fertile Barlavento district east of La Guayra, a region watered by two rivers, Rio Tuy and Rio Chico, and with the following outlets; Rio Chico (which gives its name to the most ordinary of sorts), Higuerote, and Capaya. The plantations hard on the mountainous coastal slopes produce a very fine bean, of equal value with the Puerto Cabello.

4. C a r u p a n o cacao, a sound Venezuelan medium sort, generally coming into use uncoloured; the arrivals from the easterly harbour Rio Caribe also belong to this sort, and also the cacaos of Irapa, Guiria, and Cano Colorado, often shipped from the port of Trinidad lying opposite.

From A n g o s t u r a (Ciudad Bolivar) on the Orinoco and San Fernando on the Apure, only very insignificant quantities arrive.

They speak of a Christmas and a Summer (June 21st) harvest in Venezuela; but the first four months of the year are generally the most productive. The total produce of Venezuela amounts to about 16,000 tons, of which as export there fall to

48

La Guayra	about 8,000 tons.
Puerto Cabello	about 3,000 tons.
Carupano	about 4,500 tons.
Maracabio and via Trinidad	about 500 tons.

C. The Antilles.

Trinidad produces a cacao which on many plantations, or estates, as they are called, receives preparation at the hands of experts, and is very highly esteemed in commerce, and especially in England and France. The best and generally slightly coloured sorts are specified as "Plantation", the medium "Estates", after the English name, and the inferior "Fair Trinidad shipping cacao The bean "Trinidad criollo" is oval, yet not so rounded as the Venezuelan; its kernel is for the most part dark-coloured, still brown in the better varieties, but inky black among the inferior. It is customary in Trinidad to trade the cacaos as prime specimens and to assign to them the name of a species which not infrequently furnishes no true indication of their origin. "Soconusco" and "San Antonio" are particularly high-sounding; mention can further be made of "Montserrat", "La Gloria", "Maraval", "Belle Fleur", "El Reposo" etc. Chief harvest, December to February inclusive, by-harvest May to August.

The total export from Trinidad amounts to about 22,500 tons yearly. The substantially smaller island of Grenada, also British, contributes about 6,000 tons a year to the world's supply. Owing to the prevalence of like climatic and geological conditions, the yield and quality are here the same as on the neighbouring island of Trinidad. The chief consumer of the Grenada cacaos is the Motherland, and the same holds good for the small British islands of St. Vincent, St. Lucia and Dominique, all of little import in the general trade of the world.

Martinique-and Guadeloupe-cacaos, hailing from the French islands so named, with a yearly production varying from 5,000 to 7,500 tons, only come into consideration for the consumption of the Motherland, which affords them an abatement of 50 percent in connection with the tariffs. San Domingo, the larger and eastern part of the Haiti island, already contributes about 20,000 tons yearly to the universal harvest. Especially in the last ten years has the cacao cultivation here received considerable expansion (yield 1894 2,000 tons, 1904 13,500 tons) and as vast suitable tracts of land are to hand, this country would justify the highest expectations, if the general political and economical relations of the double republic and a certain indolence of the planters, all small farmers, had not to be allowed for.

A methodical preparation only seldom takes place. Processes are limited to a very necessary drying, as a rule, so that the cacao, excellent in itself, takes rank among the lowest as a commercial quality. The chief gatherings occur in the months of May, June and July. The shipping ports are Puerta Plata on the north-coast, Sanchez and Sumana on the Bight of Samana, and La Romana, San Pedro de Macoris and Santo Domingo (the capital) on the south coast. Tiny Samana, situated on a small tongue of land, and so outlet for no extensive region, has given its name to Domingo cacao as a commercial sort, as from here the first shipments were dispatched.

S a n c h e z cacao, so named because Sanchez, where the transports come from the fruitful district of Cibao as far as La Vega, is the chief exporting harbour of the republic. From the same district, starting at Santiago, there is yet another line, this time running northwards to Puerto Plata on the coast. The cacao of this northerly province of Cibao is generally held in higher esteem than that coming from the southern harbours.

51

The United States, which have recently developed an interest in the land for political reasons, have been promoted to first place among its customers during the last few years; and then follow France and Germany. It can only be hoped that this influence grows, in view of the thereby doubtlessly accelerated improvements in the preparation processes. Up to the present, varieties free from blame are conspicuously rare. Uniformity as regards the weight of the sacks has not been possible, owing to the diversity of the means of transport. Districts lying along the railways, or close to the harbours, make use of 80-100 kg. sacks (about 176-220 lbs.) But where transport must be made on beasts of burden, sacks of from 65-70 kilos (143-154 lbs.) are the rule.

H a i t i cacao, coming from the Negro republic of the same name, is the most inferior of all commercial sorts, chiefly on account of the incredibly neglective preparation which it undergoes, for exceptions prove that the bean is capable of being developed into a very serviceable cacao. Beans covered with a thick gray coloured earthy crust, often even mixed with small pebbles and having a gritty, and where healthy, black-brown beaking kernel. The "Liberty and Equality" of the Negros and Mulattos in this corrupted republic are mirrored in its plantation system, the land being cultivated but little, and running almost wild. To effect a change in this state of affairs, that island law must first of all be abolished, whereby every stranger is prevented from acquiring landed estate in Haiti.

The yield, about 2,500 tons, is chiefly exported from Jérémic, then also from the harbours Cap Haitien, Port de Paix, Petit Goave, and Port au Prince. France and the United States are the principal customers. The neighbouring island of

C u b a also delivers the greater part of its cacao produce

to the United States, amounting to between 1,000 and 3,000 tons, a fact explained by geographical, political and freight considerations.

Thanks to its careful preparation, this bean, which resembles the Domingo in many respects, is preferred, and fetches a correspondingly higher price. The shipping port is Santiago de Cuba, situated in the south-eastern portion of the island.

Jamaica, with its yearly harvest of about 2,500 tons, principally attends to the wants of the Mother Country.

II. African Cacao Varieties.

Cacao cultivation in Africa is of comparatively recent date. The plantations found on the three islands San Thomé and Principe (Portuguese), and Fernando Po (Spanish), lying in the Gulf of Guinea, are the oldest. To the first-named island may be traced much of the impulse given to cacao plantation in other African districts, so rapid has been its success here, under the energetic guidance of the skilful Portuguese planter, and the yet more effective propitious climatic influences and favourable industrial conditions.

Rare sorts are nowhere to be met with, for the Forastero bean has conquered the whole of Africa. The sorts produced are accordingly rather adapted for general consumption. St. Thomas and the Gold Coast provide a third of the world's present-day cacao supply, and in the English colony especially, the geological and climatic conditions are of such a kind, that the

Gold Coast might very well become to the raw cacao market of the future what the Brazilian province, San Paulo, is now to the coffee trade.

In the middle of the "Eighties", the Swiss Missionary Society planted in the vicinity of their station, and so started the cultivation of the cacao tree now flourishing throughout the land. The first fruits came to Europe in 1891, and in 1894 already totalled 20 tons. In 1901 it was 1,000 tons, 1906 approaching 10,000 tons, and the year 1911 provided the record with about 40,000 tons. It is true that complaints were long and rightly lodged concerning the inferior quality, due to carelessness on the part of the natives in conducting the processes of preparation. But since the year 1909, there have appeared on the market side by side with the inferior and so-called current qualities, which still retains more or less of the defects of the earlier produce, another and properly fermented cacao, in no mean quantities; it is very popular in all cacao-consuming lands, and fetches from 2 to 3 shillings per cwt. more than the current qualities. All this has been achieved through intelligent and sympathetic guidance and control of the small native planter on the government's part, without resource to any large organised plantation system.

A c c r a cacao, then, as the sorts of the African Gold Coast are collectively named, also promises to be the cacao of the future, if it can maintain its quantitative and qualitative excellence. There is indeed no want of soil and adequate labour strength in that province. Apart from Accra, Addah, Axim, Cape Coast Castle, Prampram, Winebah, Saltpond, Secondi must be mentioned above all. The chief harvest is from October to February.

T o g o, the small German colony adjoining the British Gold Coast, has till now had only a yearly yield of 250 tons in a variety resembling Accra. The excellent beans prepared on the plantations fetch several shillings a cwt. more than Accra, whilst the deliveries of the natives rank below the current specimens of this sort. Its port is Lome.

L a g o s, the British Colony bordering on Dahomey and east of the Gold Coast, is watered by the Niger and possesses cacao exporting ports in Lagos, Bonni and Old Calabar, and exports about 4,000 tons of a sort resembling Accra, but nevertheless not so well prepared and so of inferior value.

The cacao plantations of the Lagos colony,—more properly known as Southern Nigeria—lie on either side of the great Niger delta, in low lying land where the climatic and geological conditions are quite different from those in the neighbouring German possession of

K a m e r o o n, in which country steep slopes and the narrow coastal strip at the foot of the Kameroon range, lofty mountains, perhaps 13,000 ft. high, constitute the cacao cultivating region. Consequently the same variety of seed, the Forastero, here produces a different kind of fruit. The Kameroon bean has its own peculiar characteristics; although there is some resemblance to that produced on the opposite islands of Fernando Po, Principe, and St. Thomas; and the milder sorts from the "Victoria" and "Moliwa" plantations often do duty as a substitute for the latter variety. There is no other bean which contains so much acid as the Kameroon, and although this statement must be modified in view of improvements in recent years, the fact prevents the largest of German colonial sorts from serving as any other than a mixing variety.

Cultivation is the rule throughout Kameroon, with the exception of Doula, and the produce of the separate plantations, such as Victoria, Bibundi, and Moliwe, Bimbia, Debundscha and so forth, all of which belong to large Berlin and Hamburg companies, is influenced and differentiated by variations in the technique of preparation. There are smooth beans with blackish-brown shells, and others of a red-brown hue and shrivelled, some with traces

55

of fruit pulp, and others again quite light-coloured, with occasional black specks resulting from a too thorough drying.

The chief gathering begins in September and ends in January. Exportation began in the year 1899 with 5 cwts. The produce in 1898 figured at 200 tons and it had in the year 1910 grown to 3,500 tons. Germany is of course the principal consumer, although England has since 1909 bought very much Kameroon cacao as St. Thomas.

Kongo is a bean resembling the finer St. Thomas, but smaller and often smoky. It comes on the market via Antwerp. Up to the present French Congo has only produced a few thousand hundredweights yearly, but the Belgian Congo Free State has managed to achieve an annual output of 900 tons towards the close of the last decade; and when this country takes the Gold Coast as model, perhaps Congo cacao will one day play an important rôle in the world of commerce.

St. Thomas the small Portuguese island lying in the Gulf of Guinea, and almost on the Equator, produces a sort which enjoys immense popularity, and especially in Germany, which traces a fourth part of its consumption back to this island. The export figures are

> 1889 2,000 tons.
> 1894 6,000 tons.
> 1899 11,500 tons.
> 1904 18,000 tons.
> 1910 38,000 tons.

These are estimates which make the Portuguese planter worthy of all respect. It is true that "Black ivory" has been utilised on a large scale, the exploiting of black labour having resulted in a boycotting of these St. Thomas sorts on the part of some English manufacturers, but less on account of harsh treatment on the plantations themselves as the manner of recruiting in Angola.

Fine Thomas is the description of those sorts which have been used in an unmixed condition owing to their indigestibility, but properly gathered and fermented. The inferior and slightly damaged cacaos picked out from these are called by the Portuguese planter "Escolas", or assorted. Yet they do not come into commerce under this designation, being mostly used for making up sample collections which illustrate the difference between these and F i n e T h o m a s The latter is traded through Lisbon "On Approval of Sample

All the St. Thomas cacao trade passes through Lisbon; for the tariff regulations of the Portuguese government make direct connection between the island and the consuming land practically impossible. France indeed chooses the route via Madeira, unloading and reloading, to avoid the additional duties. The cacao is at Lisbon stored in the two great Custom-houses there, and prepared for despatch to the respective lands. Fine St. Thomas is reshipped in the original sacks.

The samples are offered under various marks, either the initials of the planter or the name of a plantation. We mention a few of the best known; U. B., D. V., R. O., "M. Valle Flor", "Boa Entrada", "Monte Café", "Santa Catarina", "Pinheira", "Agua Izé", "Colonia Acoriana", "Queluz", "Gue Gue", "Rosema", "Pedroma", "Monte Macaco

The beans vary, as far as shell and kernel are concerned, according to the mode of preparation on the plantations and the structure of the soil from which they spring. Many which were formerly universally esteemed are now no longer preferred because the soil in the meantime has been worked out; and many are now described under different marks. Yet particular characteristics still continue; there are mild and strong sorts, smooth and shrivelled varieties which look as though they have been washed, and others black like the Cameroon bean. All are offered as Fine Thomas, and enjoy an immense popularity.

Good m e d i u m T h o m a s is the commercial designation of those cacaos hailing from small plantations which have undergone a scarcely sufficient preparation owing to the lack of proper apparatus, and which are always interspersed with black or sham beans. In so far as these are delivered from large plantations, they generally owe their origin to overripe fruit, probably overlooked in the gathering season; or fruits bitten by the rats which infest this island may also contribute such beans. Almost all these inferior cacaos are sorted in the Lisbon custom-houses, and thinned down to the quality "Medium Thomas" free from objection or "Good Medium Thomas The two months of the Summer harvest, July and August, supply a somewhat better variety of cacao, known in commerce as "Pajol", i. e. literally, "Hailing from the country", which generally fetches a rather higher price. During the Winter harvest from November to February the medium St. Thomas varieties come on the market, but not before the beginning of the year, as previous to that point of time only the regular harvest of F i n e S t. T h o m a s comes into consideration. All attempts on the part of consumers to effect an improvement in the quality of the medium varieties have unfortunately hitherto proved abortive, for they are regarded as by-produce on the larger estates, and the small ones do not possess the apparatus necessary for a thorough preparation. Then again it is seen that these inferior sorts are taken off the market at very reasonable prices.

Fernando Po, a mountainous island, situated immediately

off Cameroon, may be regarded as a source of supply for the Motherland, Spain, and only as such, for its yearly output of 2500 tons need fear no competition, thanks to the excessive tariffs laid on the produce of other lands here. The qualities here are inferior to those from St. Thomas and Cameroon, chiefly because most plantation are in the hands of blacks and consequently not well managed.

German East Africa, Madagascar, Mayotta (Comoren) and Réunion with their dwarfish yield are only worthy of passing mention.

III. Asiatic Cacao Sorts.

The only cacao plantations deserving the name on the continent of Asia are those occurring on the two islands of Ceylon and Java, both producing a sort differing entirely from the Africans, the predominant seedling here planted being the Trinidad-Criollo. The Ceylon-Java bean is, like the genuine Criollo, oval shaped, inclining to a sphere; its kernel is light brown and among the finer sorts even whitish. So both varieties are principally used for colouring and covering the cacao mass, for neither has a very pronounced flavour. The shell is light brown or reddish brown after washing, and appears free from all traces of pulp. It sits loosely on the kernel, at least in the case of the Java bean, and is consequently often met with broken.

Ceylon, with the shipping port of Colombo, produces in a good year from 3,500 to 4,000 tons, about two-thirds of which are traded through London. Direct shipments to Germany have recently been more and more frequent; Australia also claims consideration as a consuming land.

The different sorts, or rather qualities, for a very careful preparation ensures the excellence of the goods, go under the description fine, or medium, or ordinary, and occasionally are utilised as typical examples. The better sorts come exclusively from plantations, and the ordinary are the result of native enterprise.

J a v a also produces a large quantity, the cacao here being chiefly planted on the north side of this long, narrow island. More than a half is exported from the port of Samarang, then follow Batavia, Soerabaja and a few minor places, with a total output of about 2,500 tons. The larger proportion of this cacao is sold in the markets of Amsterdam and Rotterdam to Dutch merchants, who pass it on to other consuming countries. England, North America, Australia, China and the Philippines are the chief customers.

Those sorts coming from the neighbouring islands of Celebes, Timor, Bali, Amboina and Lombok may also be considered as sub-classes of the Java; but they do not total more than 75 tons.

IV. Australian Cacao Sorts.

Cacao plantation in Australia is still in its early stages. Most progressive is

S a m o a, which has increased its 1900 export of 30 cwt. to 200 tons at the present time, among which right excellent qualities occur, culled from Criollo trees. The deteriorated Forastero has also recently been planted, which we must allow to be more fruitful and less dependent on careful nursing. The Samoa Criollo bean resembles the large fine Ceylon variety, except that it has a more pronounced flavour.

New Guinea and Bismarck-Archipelagoes can only claim casual mention as experimentally interested in cacao cultivation.

g) The Trade in Cacao and the Consumption of Cacao Products; Statistics.

Although cacao and cacao products have always been held in the highest esteem, ever since they first became known in Europe, yet price considerations long prevented them from enjoying the same widespread popularity among the lower classes as tea and coffee. Thanks, however, to the improved means of transport established in the course of the last fifty years, which has cheapened all exotic produce, the demand for these wares has of late been more frequent and urgent, and is reflected in the constantly increasing influx of cacao on the European markets and the systematic opening out of new regions to the raw material, just as corresponding extensions in the factory world contribute towards a reduction in the cost of the products. Hence cacao may now be described as a luxury within the reach of everyman. Its diffusion among all grades of the population may be regarded as a great blessing, for in it has arisen a new [Transcriber's Note: a line is missing here] merely a stimulant, like tea or coffee, but a beverage in the proper sense of the term, analytically so established.

It will accordingly prove of interest to glance through the returns in connection with the trade in these goods, their importation and exportation, commercial values of the same, and the relative consumption of cacao, tea and coffee.

Such figures are always at hand. The surprisingly rapid growth of the cacao cultivation, and the manufacture of cacao products, is e. g. at once apparent in statistics furnished by the French government. In 1857 the number of 5,304,207 kilos of beans were consumed there. The importations of the year 1895, on the other hand, amounted to 32,814,724 kilos, having in the space of 38 years increased more than sixfold. Of this quantity, almost the half, comprising about 15,234,163 kilos, is disposed of retail.

Turning to the trade in Germany, the cacao industry here and its consumption,[19] we are again greeted with cheery prospects. According to the official inquiry, German trade in Cacao products for the years 1907-1910 is shown in the following table:

Table 1.

No. on offic. statistics	Description	Imports to Germany				Expo	
						Duty Free	
		1907	1908	1909	1910	1910	1
63	Cacao Bean raw	345154	343519	407248	439413	—	:
64	Cacao Shell whole	55	1	6	6	—	1:
168	Cacao Butter Cacao Oil	243	106	208	263	22223	2(
203a	Cacao Mass, Ground Cacao shells	165	1196	128	58	125	:
203b	Cacao Powder	6792	8148	6497	6446	2599	:
204a	Chocolate & Chocolate Equivalents	11636	10050	12197	15183	1513	!
	Products from Cacao Mass, Cacao						

| 204b | Powder, Chocolate and Chocolate Equivalents, Acorn, and Oat cacaos | 1239 | 1281 | 1258 | 1140 | 2027 | ' |

The year 1910 brought a total import of 878,413 cwts. of raw cacao, thus overtopping the figures of the previous year, which had created a record with 814,496 cwts., by 64,330 cwts.

Coming to the geographical distribution, we find that they were imported into Germany in the following proportions, namely:

		1910	1909	Comparison with previous years
British West Africa	cwts.	206 180	189 686	+ 6 494
Port. West Africa (St. Thomas etc.)	"	239 756	181 230	+58 526
Brazil (Bahia)	"	128 760	137 396	- 8 636
Ecuador (Guayaquil)	"	97 454	101 038	- 3 584
Dominican Republic (Samana)	"	64 932	66 210	- 1 278
The Rest of British America	"	21 266	40 658	- 508
Venezuela	"	40 068	36 002	-44 26
Cameroon	"	20 426	22 026	- 1 420
Ceylon	"	15 892	12 488	- 3 402
East Indies (Dutch)	"	8 802	6 772	- 2 030

Cuba	"	2 610	3 066	- 456
Haiti	"	3 676	2 614	- 1 562
Samoa	"	3 216	2 230	- 314
Togo	"	564	250	- 314

These figures, which we quote from the Thirty First Year's Report of the Association of German Chocolate Makers, speak volumes for the recent development of the cacao trade. It is interesting, in view of recent occurrences, to note the quantities despatched from the various places. The importations from St. Thomas, for instance, show a striking increase. They stand at the head of the raw cacao products coming into Germany, with 239,756 cwts., and have pushed Accras down to second place, this variety having failed to maintain its 1909 lead, for 1910 did not add more than 6,496 cwts. to its previous total of 199,686 cwts. Bahias came third, then as now, with 128,760 cwts. This order has not always remained constant, but has suffered considerable deviations in progressive years. We give below a table showing the chief cacao producing lands and their imports into Germany between 1900 and 1908.

Table 2. **Imports in Germany in tons.**

	1900	1901	1902	1903	1904	1905	19(
Brit. West Africa Gold Coast (Accra)	— —	— —	559·1	935·2	1580·9	2775·9	404
Portuguese West Africa (St. Thomas)	2501·6	3116·0	4069·2	3878·8	4526·6	4259·3	496
Brazil (Bahia)	3776·8	3239·0	3125·5	2599·8	4130·4	4506·4	610

Ecuador (Guaquil)	5397·9	4744·8	4723·6	5092·7	5689·8	5350·3	469
Dominican Republic (Samana)	586·1	1853·0	2448·8	3116·0	4562·4	4514·1	566
Rest of British North America	1436·9	1195·6	1544·7	1292·3	1851·5	2009·0	250
Venezuela (Caracas)	1158·5	956·6	893·2	829·4	1280·3	1380·9	168
Cameroon	— —	190·9	361·5	470·7	647·5	839·4	119
Ceylon	— —	107·4	344·9	350·1	497·7	589·3	588
East Indies (Dutch)	— —	— —	— —	— —	— —	— —	— -
Cuba	— —	299·8	345·3	144·7	189·0	195·6	— -
Samoa	— —	— —	— —	101·3	203·8	140·0	— -
Columbia	— —	112·6	104·3	52·6	— —	— —	— -
Togo	— —	— —	— —	— —	3·7	6·0	— -
via The Netherlands	122·1	363·9	357·6	60·9	— —	— —	— -
via Portugal (probably Thomas)	988·1	1311·4	1349·1	2447·7	1734·9	2853·4	271
Haiti	1796·0	340·4	In consequence of tariff struggle	— —	— —	— —	— -

The consumption of cacao in other civilised countries shows a corresponding increase, although with occasional divergencies and astounding relapses. We give the following table (3) to indicate its progress between the years 1901 and 1908, and to facilitate comparison.

It must be borne in mind, when making use of this table (specially in connection with Germany) that the falling off in the years 1907-8 is to be attributed to the abnormally bad harvests and consequent increase in prices.

T a b l e 3. **Import or Consumption in the Various Lands in tons.**

	1901	1902	1903	1904
The United States of North America	2066595·8	2312072·8	2850808·2	3216415·6
Germany	1841000·0	2060170·0	2163440·0	2710140·0
France	1791650·0	1934300·0	2074150·0	2179450·0
England	1890800·0	2038600·0	1868119·2	2054250·4
Holland	1437300·0	1466627·4	1073047·4	1218440·0
Spain	593107·7	925997·6	602675·2	581635·9
Switzerland	436330·0	570700·0	585650·0	683910·0
Belgium	186548·7	227763·3	276779·1	279200·8
Austria-Hungary	168650·0	182010·0	203460·0	251010·0
Russia	—	—	190068·0	205570·0
	1901	**1902**	**1903**	**1904**
The United States of North America	3523164·5	3794857·5	3752650·5	4261529·3
Germany	2963310·0	3526050·0	3451540·0	3435190·0
France	2174760·0	2340380·0	2318030·0	2044450·0
England	2119071·2	2013204·0	2015947·2	2105152·0
Holland	1073740·0	1122400·0	1221924·9	1582100·0
Spain	610171·2	563682·1	562823·9	658011·3

Switzerland	521840·0	646690·0	712420·0	582050·0
Belgium	301899·7	386168·6	325396·7	455408·1
Austria-Hungary	256850·0	331280·0	347170·0	370730·0
Russia	222768·0	267094·0	247338·0	258806·0

The relative consumption of coffee, tea and cacao has also inclined in favour of the latter as far as Germany is concerned. According to the 19th. Report of the Association of German Chocolate Makers, No. 7, the imports which passed through the custom-houses of that country, and intended for consumption, figured at the following in tons; though in this connection it is as well to remember that the German ton is about 50 lbs. less than the English.

	Coffee	Cacao	Tea
	(raw in bean)	(raw in bean)	
1886	12360·5	3686·7	1618·5
1887	101833·4	4295·0	1760·0
1888	114658·1	4979·8	1778·4
1889	113228·5	5565·1	1875·0
1890	118126·3	6246·5	1995·0
1891	125611·2	7087·0	2221·0
1892	122031·9	7460·9	2479·0
1893	122190·5	7960·9	2676·0
1894	122357·5	8319·9	2840·0
1895	122390·2	9950·9	2544·0
1896	129896·6	12209·5	2471·0
1897	136395·0	14692·5	2852·0
1898	153270·4	15464·9	3661·9

From the above columns it will be seen that the importation of coffee has only increased 24 percent, that of tea 125 percent, but that of cacao at the surprising rate of 330 percent. A comparison of the totals for coffee, tea and cacao in the years 1886, 1898 & 1906 will make the proportions still more evident.

	1886	1898	1906
Coffee	96·0%	89·0%	82·6%
Cacao	2·8%	8·9%	15·6%
Tea	1·2%	2·1%	1·8%
Total	100·0%	100·0%	100·0%

So that whilst in the year 1886 thirty-five times as much coffee as cacao found its way into Germany, the imports for 1898 were ten, and in 1906 only five and a half times greater in the case of the first named article. It follows that there has been a corresponding increase as regards cacao consumption in Germany. A momentary survey of the graphs in Fig. 5, which we owe to the kindness of Herr Greiert, Managing Director of the Association of German Chocolate Manufacturers, will make this clear to the reader; and the diagram there illustrates the relative growth of cacao consumption in Germany, when compared with other countries. On calculating the quantity of cacao consumed per head of the population, we get a graph (fig. 6) which puts the rapid increases in this direction at a glance.

Ausgestellt vom Verband deutscher Chocoladefabrikanten.
Sitz Dresden

Verbrauch von Rohkakao
1896 - 1901
in Frankreich. Grossbrittanien. Holland
den Verein. Staaten v. N.A. und Deutschland
in 1000 Dz. (100 kg).

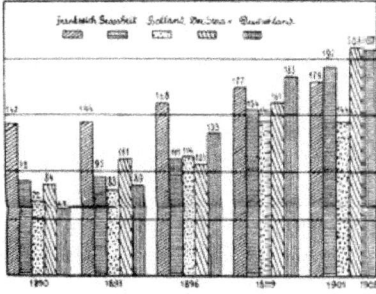

Prozentuale-Steigerung
des durchschnittl. Verbrauchs
von Kakao (in Bohnen) Kaffee u. Tee
in Deutschland verglichen mit dem
Stande von 1840.

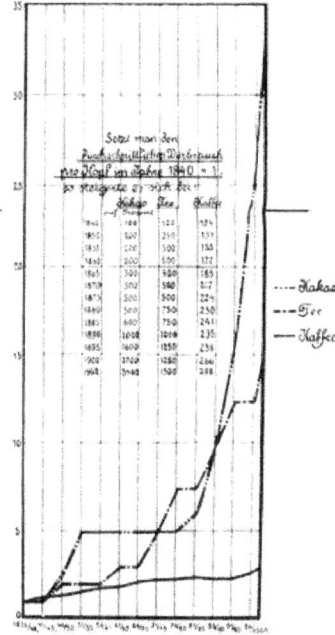

Einfuhr von Rohkakao über die Deutsche Zollgrenze
1883 - 1901
in Doppelzentnern.

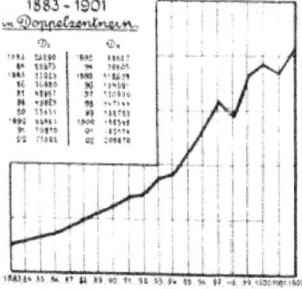

Fig. 5. (the german text is here).

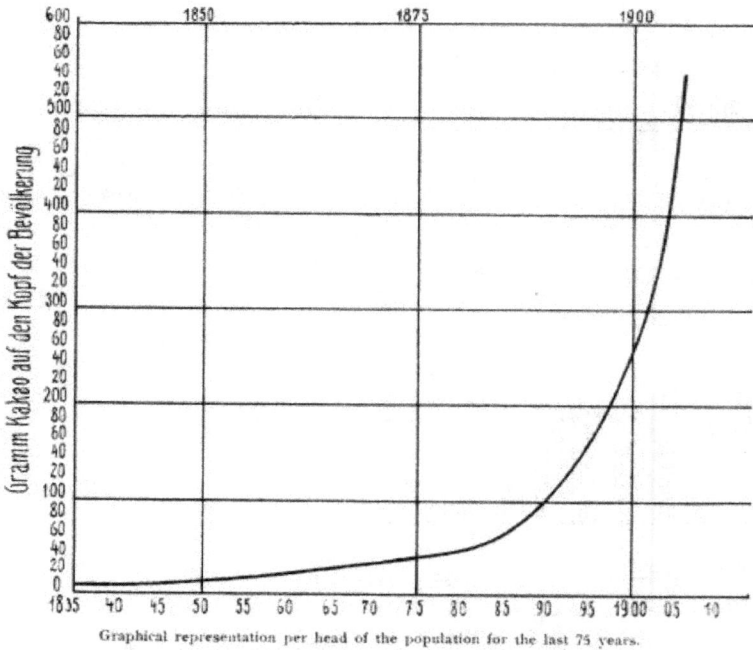

Graphical representation per head of the population for the last 75 years.

Fig. 6. Graphical representation per head of the population for the last 75 years.

The curve for the last ten years represents enormous advances, and contrasts with the more even line developed in earlier years. According to official reports, the average consumption of cacao per head between the years 1861-5 amounted to 0·03 kg. (tea 0·02 kg. and coffee 1·87 kg.) but had in 1910 risen to an average of 0·53 kg. per head.

B. Chemical Constitution of the Bean.

a) The Cacao Bean Proper.

Just as the beans of the cacao fruit are included under the botanical concept "Seed", so also their chemical constituents closely resemble those common to every other seed. There are the usual reserve stuffs inherited from the mother plant, which serve as sustenance for the yet undeveloped organs, and compare with albumen in the feathered world. Apart from the constituents incidental to all plant life at this stage, such as albumin, starch, water, fat, sugar, cellulose and mineral stuffs such as ash, the cacao seed has two other components peculiar to itself; T h e o b r o m i n e and C a c a o - r e d. We adjoin a succession of chemical determinations respecting the quantitative proportions of these substances in the seed, and think further that we may be allowed to cite the results of fore-time investigators in this sphere, especially as their work has formed the basis for all future operations, and again, in view of the doubt which still prevails in scientific circles as to the "Normal" composition of the cacao bean.

T a b l e 4.

Percentage Composition of the Hulled Bean.

Analyst	Payen[20]	Lampadius[20]	Mitscherlich[20]	
Constituents percent	Undescribed	West Indies	Guayaquil	Caraca
1. Water	10·0	3·40	5·60	—
2. Nitrogenous matter	20·0	16·70	14·39	—

3. Theobromine	2·2	—	1·20	—
4. Fat	52·0	53·10	45-49	46-49
5. Cacao-red	—	2·07	3·50	—
6. Sugar	—	—	0·60	—
7. Gum and Starch	10·0	7·75	14·30	13·5
8. Woody fibre	2·0	0·90	5·80	—
9. Ash	4·0	3·43	3·50	—

Table 5.

Constituents percent	Laube			Aldendorff	
	Caracas	Guayaquil	Trinidad	Puerto Cabello	Surina
1. Water	4·04	3·63	2·81	2·96	3·76
2. Nitrogenous matter	14·68	14·68	15·06	15·03	11·00
3. Fat	46·18	49·04	48·32	50·57	54·40
4. Starch	12·74	11·56	14·91	12·94	—
5. Other non-nitrogenous matter	18·50	12·64	12·06	11·49	28·32
6. Woody fibre	4·20	4·13	3·62	3·07	—
7. Ash	3·86	3·72	3·22	3·94	2·35

C. Heisch

Constituents percent	Granada	Bahia	Cuba	Para

1. Water	3·90	4·40	3·72	3·96
2. Nitrogenous matter	12·45	7·31	8·56	12·50
3. Fat	45·60	50·30	45·30	54·30
4. Starch	—	—	—	—
5. Other non-nitrogenous matter	35·70	35·30	39·41	26·33
6. Woody fibre	—	—	—	—
7. Ash	2·40	2·60	5·90	3·06

The analyses carried out by Zipperer in the year 1886 yielded the following results[21]:

Table 6.
A) Analysis of the Raw Shelled Bean (Kernel).

Constituents percent	Names of Sorts			
	Ariba	Machala Guayaquil	Caracas	Puerto Cabello
1. Moisture	8·35	6·33	6·50	8·40
2. Fat	50·39	52·68	50·31	53·01
3. Cacaotannic acid, sugar, decomposition products, phlobaphene	8·91	13·72	10·76	7·85
4. Theobromine	0·35	0·33	0·77	0·54
5. Starch	5·78	8·29	7·65	10·05
6. Cellulose and				

proteins	22·10	14·45	19·84	15·83
	Proteins to cellulose	Proteins to cellulose	Proteins to cellulose	Proteins to cellulose
7. In the ratio	7·3:1	5:1	6·6:1	5·3:1
8. Ash	5·12	4·17	4·17	4·32

	Surinam	Trinidad	Port au Prince	Average
1. Moisture	7·07	6·20	6·94	7·11
2. Fat	50·86	51·57	53·66	51·78
3. Cacaotannic acid, sugar, decomposition products, phlobaphene	8·31	9·46	11·39	10·02
4. Theobromine	0·50	0·40	0·32	0·45
5. Starch	6·41	11·07	8·96	8·33
6. Cellulose and proteins	24·13	18·43	15·81	18·71
	Proteins to cellulose	Proteins to cellulose	Proteins to cellulose	Proteins to cellulose
7. In the ratio	8:1	6:1	5·25:1	6·2:1
8. Ash	2·72	2·87	2·92	3·60

In addition to these, there is an exhaustive succession of analyses conducted by Ridenour,[22] which we accordingly submit as Table 8. Following Filsinger,[23] we cannot regard these analyses as an absolutely trustworthy representation of the "Normal" composition of the cacao bean, the values in starch, albumin and ash considerably deviating from all that have been established up to the present time. Among more recent researches, we cite those carried out by Matthes and Fritz Müller.[24]

Table 7.

B) Analysis of the Raw Shelled Bean (Kernel).

Constituents percent	Names of Sorts			
	Ariba	Machala Guayaquil	Caracas	Puerto Cabello
1. Moisture	8 ·52	6 ·25	7 ·48	6 ·58
2. Fat	50·07	52·09	49·24	48·40
3. Cacaotannic acid, sugar and phlobaphene	8 ·61	7 ·84	6 ·85	8 ·25
4. Theobromine	0 ·30	0 ·31	0 ·05	0 ·52
5. Starch	9 ·10	11·59	9 ·85	10·96
6. Cellulose and protein bodies	19·43	18·17	22·16	21·21
	Proteins to cellulose	Proteins to cellulose	Proteins to cellulose	Proteins to cellulose
7. In the ratio	6·5:1	6:1	7·7:1	7:1
8. Ash	3 ·89	3 ·75	3 ·92	4 ·08
	Surinam	Trinidad	Port au Prince	Average
1. Moisture	4 ·04	7 ·85	6 ·27	6 ·71
2. Fat	49·88	48·14	46·90	49·24
3. Cacaotannic acid, sugar and phlobaphene	8 ·08	7 ·69	7 ·19	7 ·78
4. Theobromine	0 ·54	0 ·42	0 ·36	0 ·43

5. Starch	10·19	8·72	12·64	10·43
6. Cellulose and protein bodies	24·39	23·06	21·82	21·43
	Proteins to cellulose	Proteins to cellulose	Proteins to cellulose	Proteins to cellulose
7. In the ratio	8:1	7·6:1	7·3:1	7·1:1
8. Ash	2 ·88	4 ·12	4 ·82	3 ·92

Table 8. **Ridenour.**

Constituents percent	Commercial Varieties				
	Bahia	Surinam	Java	Trinidad	Roasted Trinidad
1. Fat	42·10	41·03	45·50	43·66	41·89
2. Theobromine	1·08	0·93	1·16	0·85	0·93
3. Albumin	7·50	10·54	9·25	11·90	12·02
4. Glucose	1·07	1·27	1·23	1·38	1·48
5. Saccharose	0·51	0·35	0·51	0·32	0·28
6. Starch	7·53	3·61	5·17	4·98	5·70
7. Lignin	7·86	3·90	6·10	5·65	5·87
8. Cellulose	13·80	16·24	13·85	13·01	19·64
9. Extractive by difference	8·99	13·53	8·90	8·31	5·84
10. Moisture	5·96	5·55	5·12	6·34	2·63
11. Ash	3·60	3·05	3·31	3·60	3·70

Constituents percent	Commercial Varieties				
	Roasted Caracas	Granada	Tabasco	Machala	Maracaib
1. Fat	37·63	44·11	50·95	46·84	42·20

2. Theobromine	0·99	0·75	1·15	0·76	1·03
3. Albumin	12·36	9·76	7·85	12·69	11·56
4. Glucose	1·76	1·81	0·94	1·60	1·09
5. Saccharose	0·51	0·55	2·72	0·46	1·36
6. Starch	6·07	6·27	3·51	1·35	1·69
7. Lignin	9·05	5·55	6·44	5·95	7·16
8. Cellulose	11·69	13·49	12·57	11·32	17·32
9. Extractive by difference	9·22	9·72	9·26	9·02	6·79
10. Moisture	5·69	5·28	1·55	5·86	5·67
11. Ash	5·03	2·71	3·06	5·15	4·13

Table 9.

No.	Description	Moisture	Ether	Non-fatty dry substances	Mineral constituents	=ir
		%	%	%	%	
1	St. Thomas II	2·82	55·87	—	2·79	1
2	Java I	2·78	53·88	—	3·60	1
3	St. Thomas I	2·82	54·50	—	3·01	1
4	Caracas I	2·67	53·78	—	3·35	2
5	Puerto Cabello	3·34	53·29	—	3·58	1
6	Machala	2·93	53·98	—	3·34	2
7	Samana	2·94	55·28	—	3·10	1
8	Accra	2·94	53·94	—	3·19	1

B. Percentages for the non-fatty dry substa:

No.	Description					
1	St. Thomas II	—	—	41·36	6·536	4
2	Java I	—	—	43·34	8·306	3
3	St. Thomas I	—	—	42·68	7·053	4
4	Caracas I	—	—	43·55	7·692	4
5	Puerto Cabello	—	—	43·37	8·254	3
6	Machala	—	—	43·09	7·767	4
7	Samana	—	—	42·78	7·246	4
8	Accra	—	—	43·12	7·398	4

C. Percentages for the total of ash.

No.	Description					
1	St. Thomas II	—	—	—	—	7:
2	Java I	—	—	—	—	4:
3	St. Thomas I	—	—	—	—	6:
4	Caracas I	—	—	—	—	6:
5	Puerto Cabello	—	—	—	—	4:
6	Machala	—	—	—	—	6:
7	Samana	—	—	—	—	5:
8	Accra	—	—	—	—	5:

No.	Description	Alkali strength		Potassium Carbonate reckoned from Alkali strength of soluble ash	Pure ash (mineral stuffs minus Pot. Carb.)
		of the soluble ash	of the insoluble ash		
		cb. mm. Nitric			

			acid.		
1	St. Thomas II	3·6	4·8	0·25	2·54
2	Java I	10·4	6·8	0·72	2·88
3	St. Thomas I	2·6	5·0	0·18	1·83
4	Caracas I	4·6	4·8	0·32	3·03
5	Puerto Cabello	10·4	3·8	0·72	2·86
6	Machala	2·6	5·6	0·18	3·16
7	Samana	4·6	6·2	0·32	2·78
8	Accra	3·6	4·8	0·25	2·94
	B. Percentages for the non-fatty dry substances.				
1	St. Thomas II	8·7	11·6	0·60	5·94
2	Java I	24·0	15·7	1·66	6·65
3	St. Thomas I	6·1	11·7	0·42	6·63
4	Caracas I	10·6	11·0	0·73	6·96
5	Puerto Cabello	24·0	8·8	1·66	6·59
6	Machala	6·1	13·0	0·42	7·35
7	Samana	10·8	14·5	0·74	6·50
8	Accra	8·3	11·1	0·58	6·82
	C. Percentages for the total of ash.				
1	St. Thomas II	133·1	177·4	9·18	90·82
2	Java I	289·1	189·1	20·00	80·01
3	St. Thomas				

I	87·0	167·0	6·00	94·04	
4 Caracas I	137·9	143·9	9·50	90·51	
5 Puerto Cabello	290·7	106·6	20·10	79·89	
6 Machala	78·5	—	5·40	94·59	
7 Samana	149·0	200·0	10·20	89·79	
8 Accra	112·2	150·0	7·8	92·16	

No.	Description	Phosphoric acid			Silicic acid (SiO_2)	Ferric acid (Fe_2O_3)
		total	soluble in water	insoluble in water		
		%	%	%	%	%
1	St. Thomas II	1·0243	0·2474	0·7769	0·0154	0·0416
2	Java I	1·0753	0·4667	0·6086	0·0300	0·0224
3	St. Thomas I	1·1136	0·3621	0·7515	0·0122	0·0464
4	Caracas I	1·2708	0·3392	0·9316	0·0080	0·0184
5	Puerto Cabello	1·1433	0·4692	0·6741	0·0260	0·0207
6	Machala	1·2836	0·3647	0·9189	0·0116	0·0200
7	Samana	1·0881	0·3213	0·7668	0·0090	0·0560
8	Accra	1·1221	0·3672	0·3549	0·0082	0·0284

B. Percentages for the non-fatty dry substances.

1	St. Thomas II	2·4795	0·5989	1·8806	0·0373	0·1007
2	Java I	2·4790	1·0769	1·4021	0·0692	0·0517
3	St. Thomas I	2·6092	0·8484	1·7608	0·0286	0·1087
4	Caracas I	2·9180	0·7789	2·1356	0·0184	0·0422

5	Puerto Cabello	2·6361	1·0819	1·5542	0·0600	0·0477
6	Machala	2·9837	0·8481	2·1356	0·0269	0·0464
7	Samana	2·5435	0·7511	1·7934	0·0214	0·1309
8	Accra	2·6023	0·8516	1·7507	0·0191	0·0658

C. Percentages for the total of ash.

1	St. Thomas II	37·94	9·16	28·78	0·571	1·541
2	Java I	29·87	12·96	16·91	0·833	0·623
3	St. Thomas I	37·27	12·12	25·15	0·408	1·551
4	Caracas I	37·94	10·12	27·82	0·240	0·549
5	Puerto Cabello	31·94	13·11	18·83	0·727	0·578
6	Machala	38·42	10·92	27·50	0·346	0·597
7	Samana	35·12	10·37	24·75	0·295	1·806
8	Accra	35·18	11·51	23·67	0·258	0·889

T a b l e 10. **Commoner Varieties.**

Key to Column Headings

C; Moisture
D; Ether extract
E; Mineral matter
F; Potassium Carbonate reckoned on alkali soluble in water
G; Pure ash (mineral matter minus K_2CO_3)
Ha; according to König, as modified by us
Hb; as yielded by the Wender process
I; Silicic acid (SiO_2)
J; Ferric oxide (Fe_2O_3)
K; Soluble in alcohol P_2O_5

No.	Description	C %	D %	E %	F %	G %

No.	Description					
1	Superior Ariba, Summer crop	6·95	26·17	7·45	2·07	5·38
2	Machala 81%, Thomé I 19%	5·94	28·79	7·06	1·99	5·07
3	Machala 53%, Thomé I 47%	6·47	25·73	7·15	2·14	5·01
4	Cameroon	6·36	26·41	7·05	2·33	4·72
5	Thomé I 73%, Samana 27%	7·97	24·90	6·89	2·29	4·60
6	Thomé II 60%, Samana 20%, Accra 20%	7·37	22·85	7·39	2·24	5·15
7	Accra 60%, Thomé II 40%	6·93	22·80	7·36	2·25	5·11
8	A}Same variety,	6·56	18·96	7·61	2·14	5·47
9	B}more defatted	6·06	24·75	7·16	2·01	5·15
10	C}less defatted	5·58	29·72	6·57	1·89	4·68
11	Monarch double Ariba(R. & Cie.)	7·59	14·80	8·32	2·32	6·00
12	Helios(R. & Cie.)	7·37	17·25	7·91	2·12	5·79
a	Ariba shells (R. & Cie.) very fine ground	7·17	14·00	7·40	2·20	5·20
b	germs, Ariba (R. & Cie.) very fine ground	6·64	18·02	6·93	2·43	4·50

No.	Description	Raw Fiber		I	J	K
		Ha %	Hb %	%	%	%
1	Superior Ariba, Summer crop	4·20	4·60	0·0170	0·0522	0·0605

2	Machala 81%, Thomé I 19%	5·00	5·47	0·0172	0·0373	0·0625
3	Machala 53%, Thomé I 47%	5·20	5·42	0·0186	0·0513	0·0612
4	Cameroon	4·63	4·64	0·0160	—	0·0669
5	Thomé I 73%, Samana 27%	4·20	4·38	0·0167	0·0753	0·0690
6	Thomé II 60%, Samana 20%, Accra 20%	4·23	5·00	0·0208	0·0678	0·0726
7	Accra 60%, Thomé II 40%	4·06	4·40	0·0198	0·0545	0·0766
8	A} Same variety,	4·00	5·24	0·0390	—	—
9	B} more defatted	3·58	4·61	—	—	—
10	C} less defatted	3·20	4·42	—	—	—
11	Monarch double Ariba(R. & Cie.)	6·90	—	0·0420	—	0·0877
12	Helios(R. & Cie.)	6·40	—	0·0340	0·0400	0·0930
a	Ariba shells (R. & Cie.) very fine ground	7·49	—	0·2976	—	0·0383
b	germs, Ariba (R. & Cie.) very fine ground	7·42	—	—	—	0·0587

Table 11. **Analysis of Cacao.**
Dry product, defatted and free from alkali.

C; Defatted and alkali-free dry products
D; Pure ash (mineral substances less K_2CO_3)
E; Ash insoluble in water
F; Alkalinity of the insoluble ash Nitric acid
Ga; total
Gb; soluble in water
Gc; insoluble in water
H; Silicic acid (SiO_2)
I; Ferric oxide (Fe_2O_3)
J; P_3O_5 soluble in alcohol
K; after König (modified)
L; as yielded by the Weender process

No.	Description	C %	D %	E %	F ccm
1	Thomé II	41·06	6·186	4·725	11·7
2	Java I	42·62	6·757	3·754	15·9
3	Thomé I	42·50	6·659	4·353	11·8
4	Caracas I	43·23	7·010	4·904	11·1
5	Puerto-Cabello	42·65	6·706	4·056	8·9
6	Machala	42·91	7·365	4·894	13·1
7	Samana	42·46	6·548	4·357	14·6
8	Accra	42·87	6·858	4·292	11·2
9	Ariba	64·81	8·301	—	—
10	Machala + Thomé I	63·28	8·013	—	—
11	Thomé + Machala	66·66	7·517	—	—
12	Cameroon	64·90	7·273	—	—
13	Thomé I + Samana	64·84	7·095	—	—
14	Thomé II,				

No.	Description				
	Samana + Accra.	67·54	7·625	—	—
15	Accra + Thomé II	68·02	7·513	—	—
16	A	72·34	7·561	—	—
17	B	67·18	7·666	—	—
18	C	62·80	7·452	—	—
19	Monarch Ariba (R. & Cie.)	75·29	7·969	—	—
20	Helios Ariba (R. & Cie.)	73·39	8·880	—	—
a	Shells	76·63	6·786	—	—
b	Germs	72·91	6·173	—	—

No.	Description	Phosphoric Acid (P_2O_5)				
		Ga	Gb	Gc	H	I
		%	%	%	%	%
1	Thomé II	2·4947	0·6025	1·8922	0·0375	0·1013
2	Java I	2·5229	1·0950	1·4279	0·0704	0·0525
3	Thomé I	2·6202	0·8520	1·7682	0·0287	0·1091
4	Caracas I	2·9391	0·7846	1·1545	0·0185	0·0425
5	Puerto-Cabello	2·6807	1·1001	1·5806	0·0610	0·0480
6	Machala	2·9914	0·8499	2·1414	0·0270	0·0466
7	Samana	2·5626	0·7802	1·7824	0·0212	0·1319
8	Accra	2·6175	0·8565	1·7610	0·0191	0·0662
			Raw fibre			
		J	K	L		
9	Ariba	0·0933	6·48	7·10	0·0262	0·0806
10	Machala + Thomé I	0·0934	7·90	8·64	0·0272	0·0590

11	Thomé + Machala	0·0919	7·80	8·13	0·0280	0·0770
12	Cameroon	0·1030	7·13	7·15	0·0246	—
13	Thomé I + Samana	0·1064	6·48	6·75	0·0258	0·1162
14	Thomé II, Samana + Accra.	0·1075	6·27	7·40	0·0308	0·1004
15	Accra + Thomé II	0·1126	5·97	6·47	0·0290	0·0801
16	A	—	5·53	7·24	—	—
17	B	—	5·33	6·87	—	—
18	C	—	5·10	7·04	—	—
19	Monarch Ariba (R. & Cie.)	0·1165	9·16	—	0·0558	—
20	Helios Ariba (R. & Cie.)	0·1266	8·72	—	0·0446	—
a	Shells	0·0499	9·77	—	0·3884	0·0545
b	Germs	0·0805	10·18	—	—	—

1) See Table 9 A and Table 10.

The foregoing tables provide us with a general idea of the chemical constituents of the cacao bean, but their distinctive properties, both chemical and physical, still remain to be defined, with which we accordingly proceed, as such data will on the one hand enable us to grasp how loss may be avoided in the manufacture of cacao and chocolate wares, and at the same time render intelligible familiar processes connected therewith.

As we have seen, the following substances occur in cacao in varying amounts:

1. Water.
2. Fat.
3. Cacao-red.
4. Theobromine.
5. Albumen.
6. Starch.
7. Cellular tissue or cellulose.
8. Small percentages of grape and cane sugar.
9. Mineral or ash stuffs.

Like the majority of plants and plant products, the cacao bean consists of vesicles or cells, closed on all sides and arranged in a series of layers. They are constructed of cellular tissue or cellulose, and contain fat, albumen, water, starch, theobromine, cacao pigment, besides sugar and salts in inferior quantities.

1. Water or Moisture.

There is present in the bean from 6 to 8 percent of water, a factor which bodes well for the proper germination of the seed, as when this latter is deprived of moisture, e. g. in the course of a too thorough drying, it speedily decays. Water is still evident in small quantities even in the largest and almost withered beans, as will be seen on comparison of the foregoing analyses.

2. Fat.

As a constituent at the expense of which respiration is effected, fat remains one of the most important resources of plant. It has a twofold excellence in this connection, and firstly as a highly calorifacient and carboniferous substance, and again because such a reserve enables the living organism to oxidise with particular ease, wherefore it is found accumulated in somewhat significant measure in the majority of seeds. When seen under the microscope it appears either as round coherent masses, or as crystalline aggregates clearly distinguishable from the rest of the cell contents on treatment with a solution of osmic acid. The fat in the cacao bean usually amounts to from 50-56 percent, or one half of the total weight of the shelled beans; the shell also contains from 4 to 5 percent of fat.[25] The unfermented bean has frequently, in addition to its bitter taste, a most unpleasant flavour, attributable to the rancidity of its fatty contents.

The raw bean contains rather more fat than the roasted bean, for whilst the one averages from 50 to 55 percent, there is seldom more than 48-52 percent in the other. The cause of this phenomenon may be connected with the enrichment of the shells in fat, and in some instances, as when the beans are over-roasted, is to be ascribed to the chemical change which the play of burning heat on fatty bodies involves, when a destructive decomposition of the whole ensues, with formations of acroleine. Chemically considered, cacao butter consists of a mixture of so-called esters, or compounds connected with ether, such as the glycerides of fatty acids, and contains, in addition to stearine, palmatine, and laurine[26], the glyceride of arachidic acid. It was also formerly supposed that formic, acetic and butyric acids were among the constituents of this ingredient, but the view has been proved erroneous by Lewkowitsch[27]; similarly, the presence of theobromic acid alleged by Kingzett[28] has been called into question by Graf.[29]

Cacao butter is a fairly firm fat of pleasant taste and smell, which varies in colour between yellowish white and yellow. When freshly expressed, it has frequently a brownish shade, passing after a short time into a pale yellow, and turning almost white on long keeping. The brown colour is due to pigment in suspension, which becomes sediment in the course of melting, when the butter asumes a normal colour, referrible to pigment dissolved in the butter oils, and secondarily to a dissolution of the products of roasting in these liquids, rather than to any matter in suspension. The pleasant smell and taste of cacao butter is probably closely allied to the dissolved substances mentioned.

The fat extracted from cacao by solvents differs essentially from that obtained by hydraulic pressure, a fact overlooked in some of even the most recent experiments, and which therefore cannot be too strongly emphasised. Extracted fat is yellowish white, sometimes approximating to grey, and after having been kept a long time, the whole becomes tinged with an actual whiteness, which first attacks the outer surface, and then rapidly progresses towards the centre in concentric paths, and which is a sign of rancidity. Its fracture is partly granular, the smell is not so pronounced as that of expressed fat, being even unpleasant at times, as in the case of faulty wares (but compare page), and it has a keen taste. Cacao butter does not, as is generally supposed, keep better than other vegetable fats, but is equally liable to become rancid, as Lewkowitsch[30] demonstrates. By rancidity is denoted that state of offensive taste and smell acquired by fatty substances on longer or shorter keeping and especially when they are not properly stored. What chemical re-arrangements of the respective constituents this state presupposes is very questionable; though it appears from the experiments of Lewkowitsch[30] and others[31] that the formation of acids does not play as prominent a part as the experimenter is inclined to think, notwithstanding the marked increase in quantity which may occur. The primary cause of rancidity will rather be found in the oxidation products of the glycerine contained in all fats.

The specific gravity of cacao butter varies considerably, according as it has been expressed or extracted by means of solvents. White[32] asserts that it can only be determined when the liquefied oil has been solidified several days. According to Rammsberger the specific gravity of expressed butter is 0·85; that of butter extracted by treatment with ether figures at 0·958. Hager gives the normal specific gravity of fresh cacao butter at 15° C. as from 0·95 to 0·952; stale butter 0·945 to 0·946, and the same figures have been confirmed by other investigations, though Dietricht gives 0·98 to 0·981 at 100° C. The melting point is generally regarded as 33° C.; there is in this respect, however, a great difference between the two descriptions of fat. Expressed fat which has been kept for some length of time melts between 34° C. and 35° C., and these figures remain constant, so that it is advisable to read the melting point of fat which has been in store some time rather than that of the fresh pressed product, and take this as a standard. All other fat shows a lower melting point.

As the melting point of freshly melted cacao butter shows considerable fluctuation, the liquid fat must be kept in darkness and cooled with ice for about a week, and the reading should not be taken before the expiration of this time, as only then is it possible to obtain any definite and final result.

Experiments on the melting point of cacao butter as carried out by Zipperer under special conditions yielded the following values; cf. also Table 12.

Kind of bean	Melting point raw	Centigrade roasted
Machala Guayaquil	34·5	34·0

Caracas	33·5	34·0
Ariba	33·75	31·5
Port au Prince	34·25	33·8
Puerto Cabello	33·50	33·0
Surinam	34·20	34·0
Trinidad	34·00	34·0

White and Oldham[33] give the following melting points:

Guayaquil 33·6-33·9

Granada 33·0-33·3

Trinidad 31·5-32·5

Caracas 33·0-33·6

Ceylon 33·9-34·2

Filsinger and Henking found[34]:

Cauca 32·1-32·4

Bahia 32·7-33·4

Porto Plata 33·1-33·6

These results vary somewhat, but the differences are to be ascribed to the methods employed and to the manner in which the observations of different experimenters are carried out. Generally it may be taken that the melting point should not be under 3° or over 35°C. The fat solidifies between 21·5° and 23° C. (solidifying point). The fatty acids from the fat melt at 48°-52° C.; they begin to solidify at 45° C., the solidifying ending generally at 51°-52° C. (see table 12).

Adulteration of cacao fat, as many experiments have shown, cannot be detected simply by deflections in the melting point. Björklund's ether test,[35] which is very suitable for the detection of an admixture of extraneous substances like tallow, wax and paraffin, is carried out as described in paragraph....

Cacao fat, like all other fats, is saponified by alkalis, that is to say, forms a soap or a chemical compound of the fatty acids with alkalis such as potash, soda, ammonia etc. On the addition of a mineral acid to the soap a salt of the mineral acid and alkali is formed, with the separation of the fatty acid. The fatty acids are of two kinds:

1. The volatile acids or those which are volatile at 100°-110° C. or more easily with steam than other vapours. These usually exist only in very small quantity in cacao fat but may considerably increase in amount in the fat obtained from imperfectly fermented beans.[36]

2. The solid fatty acids are such as are fixed, and do not act in the manner above mentioned: cacao butter consists chiefly of the glycerides of these acids.

Björklund's tests will only detect, as has been stated, admixtures of wax, paraffin, tallow and bodies of a relatively high melting point. Another method must therefore be adopted to detect fat of low melting points, as cocoa-nut fat, or liquid oils like cotton seed and sesame oils. The methods in use in connection with cacao butter are the determination of the iodine, saponification and acid values finding the melting point of the fatty acids, the Reichert-Meissl number, and by means of Zeiss' butyro-refractometer, its refractive index.

The iodine value indicates the amount of iodine percent absorbed by the fat, and is accordingly a measure of the unsaturated fatty acids. As these latter differ in amount in vegetable and animal fats, though constant for each separate kind, it is possible by means of this iodine value to recognise a genuine cacao fat and to detect adulteration. The determination of the iodine value is carried out by Hulbl's[37] method, and according to Filsinger,[38] it is advisable to let the iodine solution act on the fat for from ten to twelve hours in diffused daylight. Before determining the iodine value in cacao fat, says Welmans[39] this substance should be dried at from 100-105°C. to expel the acroleine produced by too high roasting, at the same time avoiding too high a temperature, as acroleine can then be very easily reproduced. Filsinger has determined the iodine value of many varieties of cacao butter with the following results:

Kind	Iodine value
Cauca	36·2-36·7
Bahia	36·8-37·1
Porto Plata	36·6-36·9
Ariba	35·1-36·8

Genuine cacao butter shows an average iodine value of from 33-37·5.[40]

The saponification value or Köttstorfer's number[41] expresses the number of milligrammes of potassium hydrate required for the complete saponification of 1 gramme of fat, or in other words, the amount of potassium hydrate necessary to the saponification of the fat in thents percent. Filsinger[42] gives the amount as between 192 and 202 in genuine cacao butter, although it usually fluctuates between 194 and 195. Its determination is the means of detecting adulterations with cocoa-nut butter and its preparations.

The determination of the a c i d value has lately become of importance, especially since the introduction of the so-called Dutch Ha cacao or shell butter, which is obtained from cacao refuse and is often rancid. This value or number expresses the amount of potassium hydrate necessary to neutralise the free fatty acids in 1 gramme of fat, and it is therefore a measure of the amount of free fatty acid. As this constant has been variously stated, according to the methods adopted (Burstyn, Merz), the fact must be taken into account when comparing the literature on the subject. As the constants have been determined by two different methods (Merz, Burstyn), this must be taken into consideration when comparing the various data on the acid value of fats. Whilst the "Vereinbarungen" (No. 1, 1897) in a chapter on "Food Fats and Oils" still recognise two distinct methods in the determination of free fatty acids, as well as two different ways of recording the results (degree of acidity and free acid, calculated on the oily acids) there occurs in the supplement to the recent margarine code for Germany issued by the Chancellor on April 1st. 1898, entitled "Instructions for chemical research in fats and cheeses" under c) a dictum that there is only one absolute and precise procedure in the "Determination of free fatty acids (degree of acidity) These calculations are based on the Burstyn method, which we accordingly annex, more especially as it is now in universal use. It should be observed that the method of preparation and the age of the beans, as well as that of the fat all tend to increase the acid value.

The Reichert Meissl value expresses the percentage value of the volatile fatty acids present in the fat; as already mentioned, they amount to 1·6 ccm, in cacao fat extracted by solvents. Milk chocolate, says Welmans, yields a fat having a Reichert-Meissl value of 2·5, but compare page....

The determination of the r e f r a c t i v e   i n d e x in Zeiss butyrorofractometer is of value for ascertaining the purity of cacao butter, and it serves as a control on the iodine value, for according to Roques[43] the refractive index and the iodine value stand in equal relation, so that fat having a high refractive index gives a high iodine value and vice versa. The refractive index of cacao butter ranges between 1·4565-1·4578 at 40°C. corresponding to 46-47·8 on the scala of the Zeiss butyro-refractometer. The use of the latter is recommended by Filsinger as a preliminary test for cacao butter, since with a normal refraction it is not necessary to proceed further and determine the iodine, saponification and acid values, nor the melting point. In conclusion we annex table 12, where the respective constants for different varieties of cacao butter will be found tabulated.[44]

For further information on all these methods, the reader is referred to the excellent work of R. Benedict, entitled "Analysis of Fats and Waxes": VII. Edition, Berlin.

Table 12.

Physical and Chemical Analyses of the Various Kinds of Pressed Stollwerck Cacao Butter.

	Accra	Ariba	Bahia	Guayaquil	Camer
	a) Fat				
Point of refraction at 40° C	64·3	46·1	46·9	46·5	46·0
Melting Point (Polenske)[1]	33·1	33·2	31·95	32·5	33·65
Freezing Point (Polenske)	20·0	21·55	19·35	19·8	20·95
Variations[2] between Melting Point and Freezing					

	Puerto Cabello	Thomé	Trinidad	Fluctuations of Analyses Values from	mean
Point (Polenske)	13·1	11·65	12·60	12·5	12·70
Reichert-Meissl number	0·49	0·33	0·38	0·55	0·33
Polenske[2] number	0·50	0·50	0·60	0·42	0·40
Köttstorfer number	192·4	191·7	191·4	190·8	193·2
Hübl's iodine value	35·24	34·89	37·87	36·54	34·0
Bellier's reaction[4]	violet	as 1	as 1	as 1	as 1
R. Cohn's reaction[5]					
a) Fresh fat[6]	negative	"	"	"	"
b) Rancid fat	strong positive	weak positive	positive	weak positive	positiv

b) **Fatty Acids**[7]

	Puerto Cabello	Thomé	Trinidad	Fluctuations of Analyses Values from	mean
Refractive index at 40° C	34·60	34·55	34·50	34·40	33·70
Melting Point[8]	52·90	52·95	51·80	52·90	52·00
v. Hübl's iodine value	35·88	36·27	38·78	37·78	36·02

a) **Fat**

96

Point of refraction at 40° C	46·0	46·8	46·3	46·0-46·9	46·4
Melting Point (Polenske)[1]	32·7	32·95	32·9	31·95-33·65	32·9
Freezing Point (Polenske)	20·8	18·60	20·66	18·6-21·55	20·2
Variations[2] between Melting Point and Freezing Point (Polenske)	11·9	14·35	12·30	11·65-14·35	12·7
Reichert-Meissl number	0·41	0·55	0·55	0·33-0·55	0·45
Polenske[2] number	0·40	0·55	0·55	0·4-0·6	0·49
Köttstorfer number	191·6	191·7	191·5	190·8-193·2	191·8
Hübl's iodine value	32·72	37·24	33·72	32·72-37·87	35·28
Bellier's reaction[4]	as 1	as 1	as 1	—	—
R. Cohn's reaction[5]					
a) Fresh fat[6]	"	"	"	—	—
b) Rancid	opal	opal	opal		

fat | escence+ | escence+ | escence+ | — | —

b) **Fatty Acids**[7]

Refractive index at 40° C	33·50	34·70	33·50	33·5-34·7	34·18
Melting Point[8]	51·45	52·05	52·50	51·45-52·95	52·32
v. Hübl's iodine value	33·85	39·60	36·02	33·85-39·78	36·90

Remarks 1) Exact point of liquefaction difficult to observe; therefore the average of several readings must be taken.

2) Work from the Imperial Office of Health 1907, 26, 444-463.

3) Work out of the Imperial Office of Health 1904, 20, 545-558.

4) Central Journal for Germany 1908, 36, 100.

5) Journal for Popular Chemistry 1907, 16, 308.

6) Obtained at the expiration of a four weeks' treatment as recommended by Erlenmeyer.

7) Non-volatile fatty acids, insoluble in water, from the determination of the Reichert-Meissl number.

8) Obtained as under a). Freezing Point in various cases, 1 to 8 equals 47·8 — Melting Point minus Freezing Point: 52·3-47·8 4·5.

We have already stated that there is also cacao fat in the shells, and though it only amounts to some four or five percent, it has long been the care of experimenters to recover and realise that little as fully as possible. It is commercially known as Dutch IIa or artificial cacao butter, and cannot be obtained like the fat of the kernel by mechanical means, but is obtained by some cheap solvent like benzene. The traces of benzene are very difficult to hide, and consequently this shell butter has little commercial value and its manufacture is unremunerative.

Filsinger[45] gives the iodine value of shell butter as higher than that of kernel butter, and fixes it between 39 and 40: its acid value, especially if the fat is rancid, can reach 50-60° Burstyn, i. e. 50 to 60 ccm. normal alkali for 100 grammes of fat.[46] If the free acid of shell butter be counteracted with sodium or magnesium carbonate, the neutral fat then has the normal iodine value of pure cacao butter, namely 36·5. In a sample giving an abnormally high iodine value it is always necessary to determine the acid value, and if the latter be too high, the fatty acids must be removed, when if the sample be unadulterated, the normal iodine value will be obtained. It may be noted in passing that the high acid values occurring in shell butter may be due in part to the acidity of the benzene employed as a solvent.

Cacao butter has a considerable commercial value, and is consequently liable to adulteration with many inferior fats of vegetable origin. Among these are especially beef and mutton tallow, the purified fatty acids of palm-nut oil, wax, paraffin, stearic acid, dicka fat (nucoa butter, possibly) and cocoa-nut fat, as well as the numerous preparations of the last named, variously known in commerce as Mannheim cocoa-nut butter, vegetaline, lactine, finest plant butter, chocolate butter, laureol vegetable butter, palmin, kunerol etc. Other but less commoner are the sesame cotton-seed, arachidic, margarine and hazelnut oils.

For the detection of these and similar adulterates, the reactions and analytical methods described are all-sufficient. Benedict[47] discovers that the presence of wax and paraffin considerably diminishes the saponification value, cocoa, nut fat increases it and lowers the iodine value, whereas stearic acid raises the acid value.

	Melting point	Melting Point	

	Melting point °C.	of fatty acids °C.	Iodine value
Cacao butter	30-34·5	48-52	34-37·5
Oil of Almonds	—	14	93-101·9
Sesame oil	—	26-30	106·4-109
Earth-nut (Arachis) oil	—	27-31	92-101
Hazelnut oil	—	17-25	83·2-88
Cotton-seed oil	—	38-40	106-111
Oleo-margarine	32·4-32·5	42	43·8-48·5
Beef tallow	43-49	43-46	35·4-36·5
Wax	62-64	—	8·0-11
Paraffin	38-82	—	3·9-4
Stearic acid	71-71·5	—	—
Sebin	37·6-37·8	—	43·7-43·8
Cocoa-nut fat	20-28 chiefly 26·2-26·4	24-25	8-9

	Saponification value	Acid value	Refractive index in Zeiss's butyrometer
Cacao butter	192-202	9·24-17·9	46-47·8 at 40° C.
Oil of Almonds	189·5-195·4	—	64-64·8 at 25° C.
Sesame oil	187-192	—	67-69 at 25° C.
Earth-nut			65·8-67·5 at

Hazelnut oil	191·4-197·1	—	—
Cotton-seed oil	191-197	—	67·6-69·4 at 25° C.
Oleo-margarine	195-197·4	—	48·6 at 40° C.
Beef tallow	193·2-198	—	49 at 40° C.
Wax	97-107	19-21	—
Paraffin	—	—	—
Stearic acid	195-200	195-200	—
Sebin	192·4-192·6	—	—
Cocoa-nut fat	254·8-268·4	—	35·5 at 40° C.

The presence of cocoa-nut fat can also be shown by the etherification of the fatty acids with alcohol and sulphuric acid, when the characteristic odour of the ester of cocoa-nut acid occurs. Vegetable oils, such as almond, cotton-seed, arachidic, sesame and hazelnut oils, lower the melting point of the fatty acids and raise the iodine value. Sesame oil is easily detected by Baudouin's reaction, yielding a raspberry coloration whilst pure cacao butter keeps a fine yellow or dark brown. It is possible to detect the presence of so minute a quantity as 1% of sesame oil, by means of Baudouin's reaction.

The following table, containing the analytical determinations of all fatty substances which can possibly be employed in the adulteration of cacao butter, will serve to facilitate reference to this subject.

In addition to its use in the manufacture of certain cacao preparations and for lubricating parts of machinery which come into contact with the cacao etc. cacao fat is also used in perfumery and especially in pharmacy for making suppositaries, ointments, etc., but it is of no importance in

101

like ordinary butter or lard, cacao butter is not used. It has been maintained by Benedikt[48] that when in the form of chocolate it is as easily digestible in the human organism as milk fat, which is generally regarded as offering most favourable conditions for absorbtion in the intestinal canal. The digestibility of both fats varies from 92·3 to 95·38 percent, and both, in this respect, stand very near to cocoanut fat from which the solid glycerides have been removed, and to ordinary butter, the former according to Bourot and Jean.[49] being digestible to the extent of 98 and the latter 95·8 percent.

Cacao butter is obtained as a by-product in the preparation of cocoa powder and in every country where cocoa powder is produced there is always a large trade in the former article. That is, apart from Germany, especially the case in Holland, where the monthly supply to the Amsterdam market is so large that during 1899 one firm alone—Van Houten—had 855 tons for sale. The average price of late years has considerably increased, and is now about 64-73 cents per kilogramme.

3. Cacao-red or Pigment.

The majority of investigators interested in the cacao bean have assigned its peculiar aroma and taste to the cacao-red which it develops. As previously pointed out, the young fresh bean is colourless, the pigment forming later, as can be observed in many vegetable colouring materials, such as oakand cinchona-red, madder, indigo and kola-nut red (from Sterculia acuminata). As the later investigations of Hilger[50] have shown, the fresh colourless cacao bean contains a diastasic ferment, as well as a glucoside body, which C. Schweitzer[51] has termed glocoside or cacaonin. The term glucoside may be noted in passing as including

The term glucoside may be noted in passing as including those bodies, the greater number of which occur in plants, and which by treatment with alkalis, acids or ferments are split up into an indifferent body and a sugar, generally glucose. These bodies may be chemically regarded as ethyl derivatives of the respective sugars. When the ripe, white seeds are dried, the cacao-glycoside is partly decomposed by the agency of the above-mentioned diastasic ferment and formations of grape sugar, pure non-nitrogenous cacao-red, together with theobromine and coffeine ensue. These substances, and likewise a certain amount of undecomposed cacao glycoside, can all be detected in the seed, which has by this time acquired a brownish to violet colour.

The unfermented bean, according to Schweitzer, has as much as 0·6% unaltered glucoside. Fermentation produces the same effect as drying, as here again the glycerine is not completely split up, for the cacao-red, isolated in the ordinary way, consists according to Hilger of a mixture of pure non-nitrogenous cacao-red and some glycoside.

The complete decomposition of the cacao glycoside can only be effected in a chemical manner, by boiling the finely divided and defatted seeds with dilute acids, a method which has made it possible to effect an exact determination of the diureides, as the treatment with acid sets free the totality of their theobromine and coffeine.

Schweitzer regards the molecule of cacao glycoside as an ester comprised of one molecule of non-nitrogenous cacao-red, six molecules of starch-sugar and one molecule of theobromine with double-sided attachment and having the hypothetrical formula $C_{60}H_{86}O_{15}N_4$.

Before the appearance of Hilger's researches, all statements of a chemical nature respecting cacao-red related to a mixture of a pure non-nitrogenous pigment and the glycoside, which must in all cases be preliminarily obtained, before the pure pigment can be prepared. That can be done[52] by treating the roasted beans with petroleum ether, which removes the fat and part of the free theobromine then with water, to extract the remaining theobromine, coffeine, sugar and salts, and finally with alcohol, to extract the cacao-red. The alcoholic residue is then quickly dried on porous plates. The material thus obtained is a reddish brown amorphous bitter powder, which is scarcely soluble in water, easily so in alcohol or in dilute alkali, and is reprecipitated by acid from its alkaline solution. It gives a sublimate of theobromine when heated. When the substance is distilled with 5 percent of sulphuric acid, the added glycoside is completely decomposed into sugar, theobromine and the real cacao-red, which latter is represented by the formula $C_{17}H_{12}(OH)_{10}$. It appears to stand in near relation to tannin, which it resembles in yielding formic acid, acetic acid, and pyrocatechin by the action of caustic alkalis. The pure non-nitrogenous cacao-red, at present, is of exclusively scientific interest; for practical purposes only the crude cacao-red, cacao-red glycoside, as naturally existing in the bean, is of importance. The better and the more effectual the manner in which the beans have been prepared by fermentation, the more intense is the formation of the cacao red, especially its localisation in the cells and cell tissues. This is the reason that the variations in colour of different kinds of bean and the aqueous extracts which they yield are so distinct.

Especially is this noticeable in carelessly dried beans, in which the cotyledon tissue is of a dirty brown or yellow colour instead of being brown or violet; the pigment here is not restricted to separate cells but has the appearance of having penetrated into the contiguous albuminous cells. The bean contains 2·6-5 percent of the crude cacao-red; it is and is completely extracted from the bean by weak acetic acid.

The crude cacao-red can be determined quantitatively by precipitating its solution with lead acetate, decomposing the lead precipitate with sulphuretted hydrogen and evaporating the filtrate containing the cacao-red to dryness.

The aqueous extract of the beans, which contains the cacao-red, is coloured greenish brown by alkalis, red by acids; acetates give a grey to yellowish colour; tincture of iodine, stannous chloride and mercurous nitrate give a rose to brown precipitate. Iron and copper salts produce grey precipitates which gradually become brown to black. Gelatine solution, containing alum, and albumin give copious yellow precipitates.

Stains produced on linen by the colouring matter of cacao-red can be removed by treatment with hot water and finally bleaching with a solution of sulphurous acid.

4. Theobromine.

All those materials which are regarded as stimulants, like coffee, tea, cacao, tobacco etc., owe their action to peculiar nerve stimulating bodies, which are present only in small quantity in the seeds or leaves of the respective plants and are termed by chemists alkaloids and diureides.

The physiologically active constituents of tea, coffee and cacao are considered, even up to to-day, by many authors as alkaloids or organic bases and especially ranked among the xanthine or purine bases. Recent investigations, however, separate these substances from the alkaloids in the strict sense and comprise them within a particular group of urea derivatives under the designation of ureides; the ureides of derivatives under the designation of ureides; the ureides of tea, coffee and cacao representing two molecules of urea, they are to be qualified as "diureides

A bitter substance in the cacao bean had already been observed by Schrader, but Woscressensky[53] in 1841 was the first to isolate the diureide, theobromine.

Theobromine is found in the unfermented and fermented beans in two forms; as free theobromine, which has been eliminated from the glucoside by the ferment in the drying and fermenting processes, and in combination with glucose and cacao-red as a glucoside, from which it can only be separated by chemical means.

Theobromine stands in near relation to caffeine, the diureide of tea and coffee, as will be seen from their chemical formulae—in which theobromine is shown to contain one methyl group CH_3, less, its place being taken by an hydrogen atom;

$$C_5HN_2O_3(CH_3)_3,   C_5H_2N_2O_3(CH_3)_2,$$

so that in all, theobromine falls short of caffeine by only one radical. Strecker[54] was the first to show the relation between the two substances, when he succeeded in converting caffeine into theobromine by the action of methyl oxide on silver theobromine for 24 hours at 100° C. Caffeine and silver iodide are then formed and can be separated by treatment with alcohol, which dissolves the caffeine, leaving the silver iodide undissolved.

E. Fischer[55] was shown the relation of theobromine and caffeine to uric acid by artificial synthesis of both substances from derivatives of both. Fischer, starting with monomethyl pseudo-uric acid, converted it into 7-methyl uric acid by distilling it with hydrochloric acid, and afterwards, by treating the lead salt of the latter with methyl iodide and ether, produced 3-7-methyl-uric acid. That acid was converted into dimethyldioxychlor-purine by treatment with a mixture of phosphorus oxychloride and phosphoric penta-chloride, with subsequent reduction into 3-7 dimethyl-6-amino-2-oxy-purine, from which, by the action of nitrous acid with loss of the amine group, theobromine was finally obtained. The synthesis of theobromine is a brilliant exploit of Fischer's, and it is quite possible that at no distant period, when a simple and cheap method of production has been arrived at, synthetical theobromine will appear commercially as a rival of the natural product. At present there is no prospect of this being immediately realised, and cacao shells from which theobromine is now prepared are as yet in no danger of displacement by the new substitute, but still serve as a useful by-product in the manufacture of cacao.

Theobromine and caffeine, like the alkaloids or plant bases, have a distinct physiological and even toxic action if taken in too large quantities.

From the experiments of Mitscherlich it appears that theobromine has a similar action to caffeine, but is somewhat less active owing to its being less soluble in the gastric juice. Mitscherlich's experiments with frogs, pigeons and rabbits show that 0·05 grammes killed a frog in 40 hours, 0·05 grammes a pigeon in 24 hours, and 1 gramme a rabbit in less than 20 hours. Death resulted in all cases from cramping of the spinal cord, producing either convulsions or subsequent paralysis.

The results of these experiments do not detract from the nutritive value of cacao, since the human organism requires ten times as much theobromine as rabbits to exhibit the slightest toxic symptom; in cacao mass containing 1 % not mentioned in discussion; just a head's up to PP for S&R] theobromine, that would involve the consumption of 5 lbs. averdupois of chocolate at once, a practical impossibility. Similar conditions prevail in connection with the use of tea, coffee, and especially tobacco, where symptoms of poisoning have been occasionally noticed (the nicotine peril of excessive smokers) but it would seem that cacao and chocolate are the most favourably placed of these stimulants as regards such toxic action. It appears from the experiments of Albanese[56] Bondzynski, Gottlieb[57] and Rost[58] that 3 percent of the theobromine administered passed out in the urine unaltered, whilst on the other hand 20-30 percent of that decomposed in the organism is found again as monomethyl-xanthine.

The larger proportion of the monomethyl xanthine is heteroxanthine (= 7 Methyl-X) and the inferior 3 Methyl-X. The excretion of theobromine appears to be closely connected with the quantity of urine voided, which is especially increased by the administration of theobromine. Since 1890, as a result of W. v. Schröder's[59] observations in 1888, that property of theobromine has had an extended application in practical therapeutics; theobromine has been used as a diuretic in kidney diseases, and, unlike all similar medicinal agents, it exercises no influence on the heart, a circumstance which essentially increases its therapeutic value. It can be employed for medicinal purposes, either uncombined or in the form of salicylate, acetate and certain double compounds, as sodium or lithium and theobromine salicylate or acetate.

The double compounds known as diuretin, agurin and uropherin are freely soluble in water and are therefore more readily absorbed into the system than pure theobromine, which is only with difficulty soluble in water. Through the establishment of theobromine as a medicinal agent, for which we are indebted to Chr. Gram[60] and G. See,[61] cacao husks, hitherto a waste product in the manufacture of cacao, have become of value for the preparation of theobromine, in which many of the largest German chemical factories are now engaged.

Fluctuations as regards the percentage of theobromine in the beans are so extraordinary that they can only be ascribed to the lack of prescribed and definite modes of procedure in fermenting, which obviously necessitates differences in the resulting products.

Eminger found from 0·88-2·34 percent of theobromine in the examination of a rather considerable number of commercial kinds of cacao beans and in the husks 0·76 percent of the diureide: C. C. Keller[62] has also found it in the leaves and in the pericarp. Cacao contains 0·05 to 0·36 percent of caffeine.

Theobromine is a permanent white powder, appears under the magnifying glass as small, white, prismatic or granular crystals. At first it has only a slightly bitter taste, which becomes more intense when it is kept in the mouth for some length of time; and indeed, the bitter taste of the cacao bean and its preparations is mostly due to theobromine. It sublimes at 220 ° C. without melting. This phenomenon explains why the over roasted bean, that is, the kernel of beans which by accident have been heated to more than 130-150 ° C. is poorer in theobromine than the husks. When heated to 310 ° C. theobromine melts to a clear liquid which re-crystallizes on cooling.

One part of absolutely pure theobromine dissolves according to Eminger in 736·5 parts of water at 18 ° C., in 136 parts at 100 ° C. in 5399 parts alcohol (90 %) at 18 ° C. in 440 parts at boiling (90 %) point and in 818 parts of boiling absolute alcohol. It dissolves in 21000 parts of ether at 17 ° C. in 4856 parts of methyl alcohol at 18 ° C. in 58·8 parts of chloroform at 18 ° C. and in 2710 parts of boiling chloroform[63]. Theobromine is partly decomposed by strong alkalis but by cautious addition of alkalis it forms compounds with them, which, are readily dissolved by solutions of sodium salicylate, acetate or benzoate. These double compounds under the name of diuretin, agurin and uropherin have lately become of therapeutic value.[64]

Sodium silicate and more particularly trisodiumphosphate according to Brissemoret[65] are great solvents of theobromine. One and a half molecules of the latter salt can dissolve one molecule of theobromine so that in this way it is possible to prepare a solution of nearly 2 percent. Phenol also dissolves a large quantity of theobromine, according to Maupy,[66] who has utilised this property for the determination of theobromine. The defatted cacao preparation is moistened with water and extracted with a mixture consisting of 15 percent of phenol and 85 percent of chloroform.

Theobromine, like caffeine, gives the so called murexide reaction when evaporated with chlorine water—forming amalic acid—and when a watch glass previously moistened with a little fluid ammonia is held over the last few drops at the end of the operation. The residue thus obtained has a violet colour, which serves to distinguish theobromine readily from other plant bases which do not belong to the xanthine group.

Although theobromine is the most valuable constituent of cacao beans, the importance attached to a greater or lesser amount in the beans as a commercial article was formerly much exaggerated.

The investigations of Dragendorff and others have shown that the value of various stimulants like tobacco, coffee and tea, does not entirely depend on the amount of alkaloid or diureide but partly also on the joint action of all the constituents of those articles, and it is particularly the aromatic bodies which determine their commercial value. Various kinds of coffee, for example, of inferior commercial value contain considerably more caffeine than the costly Mocca beans. The highly prized Havana tobacco ranges lower than the Sumatra kinds in nicotine content, and the same conclusion with regard to cacao would probably be correct. In support of this view, attention may be directed to the following analyses performed by Wolfram.[67]

Percentage of theobromine at 100° C.

Description	% Theobromine %	
Caracas	1·63	1·11

Guayaquil (of considerably less value than the first)	In the bean	1·63	In the shells	0·97
Domingo		1·66		0·56
Bahia		1·64		0·71
Puerto Cabello (fine kind)		1·46		0·81
Tabasco		1·34		0·42
	Average	= 1·56%		= 0·76%

Excluding the theobromine in the shells which are not used in the preparation of cacao, it will be seen from the above table that the Caracas bean, which is the finest and dearest, has an amount of theobromine which is only equal to, or even a little less, than that in the inferior beans from Guayaquil and Domingo.

5. A l b u m i n.

On the presence of albuminous bodies in the cacao bean, varying between 14-15 percent, depends to a great extent its nutritive value. The albumin in plants, unfortunately, is not to hand in a form suitable for direct absorption and assimilation in the animal organism, in fact, only a fraction of it is so available. Before considering the nutritive value of the albumin of the cacao bean it will be well to give attention to the general chemical and physical properties of albumin so far as a knowledge of them will assist in the elucidation of the subsequent matter.

Albuminous bodies or proteins occur either dissolved in the sap of plants or in a solid in the protoplasm of plant cells; also in the form of granular deposits (Aleuron granules[68]). In cacao they are apparently present in the

112

three different conditions.

The term vegetable albumen, in its more restricted sense, is meant to designate a protein substance which is soluble in water and is coagulable by heat. The greater part of the proteid which exists in the seeds and sap of plants and is coagulable by heat, is not albumin but globulin, that is to say, it is insoluble in water, though dissolved by solutions of neutral salts. Whilst many protein substances in aqueous solution require a temperature of 100 ° C. before coagulating, or becoming insoluble under certain conditions, others coagulate at 65 ° C. Concentrated acetic acid dissolves all albuminous bodies with the aid of heat, concentrated nitric acid gives a yellow coloration (xantoprotein reaction). Albuminous substances are decomposed when heated to 150 ° C. developing a dark colour, swelling up and evolving an offensive smell, finally leaving behind a difficultly combustible coaly residue.

Globulins combine with aqueous solutions of alkalis such as potash, soda, ammonia etc. producing alkaline albuminates; with acids they form acid albuminates or syntonins. Both have the property in common, that whilst they are insoluble in pure water, they readily dissolve in slightly acidulated or alkaline water, as well as in weak saline solutions, and are then no longer coagulable by boiling.

Albuminous bodies are converted first into albumoses (proteoses), and then into peptons by gastric and intestinal digestion or by hydrolytic decomposition with acids or alkalis, also by the action of steam under pressure of many atmospheres, as well as by putrefaction. Albumoses, with the exception of hetero-albumose, are soluble in water. Peptons dissolve entirely and in that condition are absorbed by the animal organism.

Albumins are precipitated from their solutions by strong alcohol, and in that way Zipperer succeeded in precipitating 4·25 percent of albumin from the aqueous extract of Trinidad cacao, which corresponds to about 25 percent of the total amount of albumen in the bean.

The results of his investigation have shown that generally more soluble albumen is present in the unfermented than in the fermented bean. Consequently, it would appear that in the finer kinds of cacao beans, in which very careful fermentation has been carried out, the albumin, owing to fermentative alteration, is rendered less soluble.

The constitution of albumin is still not sufficiently known, despite the excellent experiments of E. Fischer on this subject; generally it is regarded as having the formula:

C	52·31-54·33%
H	7·13- 7·73%
N	15·49-17·60%
S	0·76- 1·55%
O	20·55-22·98%

Accepting a mean formula corresponding to the above figures as representation of the albumen (namely $C_{72}H_{112}N_{18}SO_{22}$), it becomes possible to obtain a quantitative determination of this constituent in the plants in which it is contained. There is, for instance, 16 % of nitrogen here. Starting from such a standpoint, and determining the percentage of Nitrogen contained in a plant, and multiplying by 6·25 (i.e. 16 %), the amount of albumen is obtained. For further particulars see paragraph 4. The albumen in cacao, as previously mentioned, is in the form of globulin, that is, in a less soluble form. In cacao preparations which are required for invalids, especially those with affections of the stomach, it is important to have the albumen in a more readily soluble condition. Various attempts have been made with cacao preparations to obtain that result, and later on, full illustrations and explanations will be given on this subject. First of all, however, it is desirable to consider the scientific methods employed to ascertain the relative digestibility or indigestibility of albumen.

Professor Stutzer[69] of Bonn has been engaged in determining the action of digestive ferments of the animal organism on alimentary substances, and has worked out a method by which it is possible to ascertain the proportion of albuminous substances which can be regarded as digestible.

The method depends upon the fact that salivary, gastric and intestinal digestion can be artificially imitated in the laboratory. But as the salivary secretion only digests starch and is difficult to obtain, malt diastase, which serves the same purpose, is used instead. On the other hand albuminous material is only digested by juices of the stomach and intestines as fresh obtained from the mucous membranes of the pig or ox. If we suppose an average of 16 percent of total albumen in cocoa powder, the following results would probably be given by Stutzer's method:

Of 16 % of total albumen there are on an average:

Albumen:	corresponding to percentage of the total mass:	
7·6% soluble in the stomach	47·5%	}65%
2·8% soluble in the intestines	17·5%	
5·6% insoluble	35·0%	
16·0%	100·0%	

As shown by the experiments of Forster[70] however,

artificial digestion does not correctly represent the actual consumption of nutriment in the human body. F o r s t e r's experiments, in which cacao powder was administered to healthy men, gave a much higher value, in fact, 80 percent of the nitrogenous substance was digested, against 65 percent by Stutzer's artificial method of digestion. The results obtained by artificial digestion must therefore be increased in that proportion.

6. S t a r c h .

Starch is one of the most important constituents of cacao, as on the starch taken in conjunction with the fat and albumen depends the nutritive value of the cacao bean. As previously stated, cacao starch is one of the smallest kinds which occur in the vegetable kingdom; consequently it can easily be distinguished from the starch granules of other plants. Owing to their minuteness the concentric rings showing the stratified structure of the starch granules can only be distinguished with difficulty under the microscope. Cacao starch consists usually of globular granules, generally separate, but sometimes in aggregations of two or three. The appearance under the microscope of the starch granules is clearly shown in fig 7, which represents a section of Ariba cacao enlarged 750 times.[71]

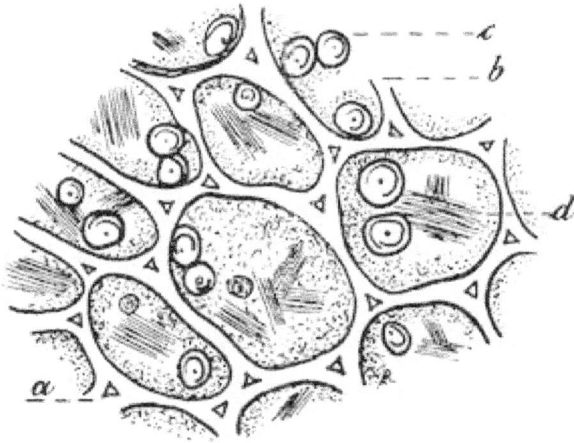

Fig. 7.

a on the above represents the intercellular spaces, b the cell walls, c the starch granules, d the fat crystals, those being the contents and structural elements of the cacao cell that the microscope will at once distinguish.

Cacao starch has the usual properties of ordinary kinds of starch, namely:

1. It is gelatinised by hot water, that is to say, the water penetrates between the layers of starch granules, separating them and causing by its penetration a swelling up of the starch whereby a transparent mass know as "starch paste" is produced. It has been supposed that cacao starch is less easily gelatinised than the starch of other plants. According to investigations of Soltsien's[72], which Zipperer unreservedly endorses, this is not the case, for under certain essential conditions, cacao starch gelatinises just as readily as other kinds of starch.

The blue coloration of starch with iodine.

This is said to take place more slowly with cacao than

with other starches, though we have always found that once the cacao starch is gelatinised, a blue coloration appears immediately on adding a sufficiently strong solution of iodine.

There are certainly other materials in the cacao bean, such as fat, which by more or less enveloping the starch, prevent access of water to the starch granules and thus hinder gelatinisation; or again, the albumen and cacao-red may exert some retarding influence on the iodine reaction, especially if the iodine solution used is very dilute Yet it is impossible to describe the reaction as slow.

According to Soltsien, if a mixture of two parts of cacao bean with one part of calcinated magnesia and water is heated, a clear-filtering decoction is obtained, which immediately assumes the blue colour on addition of iodine solution. On neutralising the filtrate with acetic acid, and adding 3-4 parts of strong alcohol, its starch is precipitated.

By boiling with dilute acids as well as by the action of ferments like the saliva, diastase etc., starch is converted into starch sugar (glucose, dextrose). The empirical formula for starch is $C_6H_{10}O_5$, that for starch sugar is $C_6H_{12}O_6$, so that in the conversion one molecule of water is introduced, wherefore its chemical nature is greatly changed, and especially in its becoming freely soluble in water. That alteration allows of starch being quantitatively determined, as the dextrose thus produced has the property of reducing an alkaline solution of copper sulphate (known as Fehling's solution, after the discoverer); that is to say, the copper sulphate is converted into insoluble red cuprous oxide. As dextrose always precipitates a definite amount of cuprous oxide, the quantity of starch present can in that way be determined.

The chemical determination of starch is only in a limited degree effectual in the recognition of an admixture of foreign starch in cacao preparations. If more than 10-15 percent of starch (calculated on the crude bean) has been found, then it must be assumed that there has been an admixture of foreign starch, but chemistry affords no means by which foreign starch can be distinguished from the genuine starch of the cacao bean. For that purpose the foreign starch must be minutely observed under the microscope, which not only serves to detect its presence but gives an approximate estimation of the amount present, and its origin. Great caution should be exercised, or the result may be easily exaggerated.

7. Cellulose or crude fibre.

We have already made the acquaintance of this material as the chief constituent of the cell walls and vascular tissues. Recent chemical investigations have shown that it consists of the anhydrides of hexose and pentose (sugar compounds) incrustated with many impurities, such as cacao-red, gum, mucilage etc. From a chemical point of view, cellulose has the same formula as starch, viz. $C_6H_{10}O_5$, or one of its multiples represented in formula. One of its chemical properties is solubility in ammonio-cupric sulphate, and affinity for alkalis such as potash, soda, ammonia, causes it to swell when they act on the cell fibres.

Weender's process[73] as worked out by Henneberg is the one usually adopted for the determination of crude fibre in plants, although recently H. Suringar, B. Tollens[74] and more particular König[75] have pointed out that in Weender's process the so-called pentosan, that is to say, the sugar-like constituent of the composition $C_5H_{10}O_5$, which comprises a not inconsiderable portion of the crude fibre, undergoes a disproportionate alteration, so that the analytical results thus obtained can by no means give an accurate representation of the amount of cellulose. The crude fibre must therefore be treated in such manner as to eliminate the pentosan. For this purpose the various methods of König, Matthes and Streitberger have been proposed, to which we shall return in Book 4. Filsinger, the meritorious experimenter on the subject of cacao, has by König's method determined the amount of crude fibre in a series of different varieties of cacao bean, and obtained the following results as regards shelled and roasted beans.

	percent
1. Puerto Cabello	5·37
2. Java	3·97
3. Ariba Guayaquil I	4·10
4. Ariba Guayaquil II	4·07
5. Machala Guayaquil I	4·43
6. Para	4·01
7. Surinam Guiana	3·01
8. Bahia	2·81
9. Grenada	3·10
10. Guatemala	3·50
11. Machala Guayaquil II	3·58
12. Caracas	3·65
13. Samana	4·58
14. St. Thomé A I	4·13
15. St. Thomé A II	2·95
16. St. Thomé B	3·15

17. Haiti	$3 \cdot 12$[76]

These new values may be provisionally regarded as normal. From these results not only can an idea of the functioning of the cacao shelling machine be obtained, but also the presence of any occasional admixture of husk in cacao preparations may be inferred, since the husk contains a great deal more crude fibre than the kernel. Therefore the determination of the crude fibre is an important item in the testing of cacao preparations, as there is no doubt that the presence of vegetable substances rich in crude fibre can be detected by the increase in the amount of cellulose.

8. Sugar and plant acids.

The presence of glucose in raw cacao beans was first pointed out by Schweitzer[77]. The sugar is formed by the action of the cacao ferment on the glucoside cacaonin during the processes of drying and fermentation. In addition to sugar, malic and tartaric acids have been observed. These substances, however, are only of interest to the plant physiologist and not to the manufacturer, so it is sufficient merely to notice them here in passing.

9. The mineral or ash constituents.

When cacao beans are ignited, the constituents of an organic nature are volatilised and only the non-volatile or inorganic constituents remain behind. These consist of potash, soda, lime, iron magnesia, combined with silicic acid, phosphoric acid, sulphuric acid and chlorine.

The amount of ash in raw and shelled cacao beans varies from 3-4 %. Tuchen[78] found 2·9-3 %, Trojanowski[79] 2·08-3·93 %, Zipperer[80] 2·7-4 %, L'Hote[81] 2·2-4 %, H. Beckurts[82]

2·20-3·75, J. Hockauf[83] 2·84-4·4 percent. Of those kinds which are now most in use, Ceylon gave 3·30 percent, Java 3·20 and Kameroon 2·95 percent. (Beckurts).

Quantitative analyses of the ash of the cacao beans have been made by several investigators, and the following table gives a series of the most complete analyses, made by R. Bensemann[84].

Table 14. **Analysis of the ash of Cacao Beans by R. Bensemann.**
The ash of the kernel free from husk dried at 100°C. contained:

Key to Column Headings

B = Maracaibo
C = Caracas
D = Trinidad
E = Machala
F = Porto Cabello
G = Mean

Insoluble respectively in dilute hydrochloric or nitric acid	B	C	D	E	F	G
a) Volatile dessicated at 100° C.	0·142	0·076	0·144	0·074	0·198	0·127
b) Fixed at red heat	0·312	1·663	0·553	0·630	1·075	0·846
Soluble in dilute hydrochloric or nitric acid:						

c) Potassium oxide K_2O	35·889	33·844	30·845	30·686	29·989	32·251
d) Sodium oxide Na_2O	0·515	0·766	1·964	4·173	3·427	2·169
e) Calcium oxide CaO	4·118	5·030	4·638	3·112	2·923	3·964
f) Magnesium oxide MgO	15·750	15·151	16·060	16·172	17·562	16·139
g) Ferric oxide Fe_2O_3	0·182	0·217	0·491	0·629	0·303	0·364
h) Aluminium oxide Al_2O_3	0·080	0·326	0·490	0·432	0·305	0·327
i) Silicic acid SiO_2	0·214	0·211	0·169	0·134	0·240	0·194
k) Phosphoric anhydride P_2O_5	27·741	29·302	28·624	37·000	35·274	31·588
l) Sulphuric anhydride SO_3	2·632	2·740	3·957	2·042	3·952	3·065
m) Chlorine Cl	0·295	0·341	0·427	0·279	0·085	0·285
n) Carbonic anhydride CO_2	10·349	8·435	8·953	2·788	3·481	6·801
o) Water H_2O	1·847	1·975	2·781	1·912	1·205	1·944
Oxygen O equivalent to chlorine	0·066	0·077	0·090	0·063	0·019	0·064

In previously describing the aleuron granules of the cacao bean it was mentioned that they contain a comparatively large globoid. According to Molisch[85], when sections are cautiously heated on platinum foil, these globules are found in the ash. From their number they give a characteristic appearance to the ash of cacao beans, and thus may serve as a good means of identifying cacao, since they can be detected in the smallest quantity of a genuine cacao preparation.

A noteworthy fact may here be mentioned, namely the presence of a rather small amount of copper in the ash of cacao beans as well as the husks. Duclaux[86] was the first to point out this fact, which several other observers, such as Skalweit[87] and Galippe[88] have also confirmed. The amount of copper in the husk varies from 0·02 to 0·025 percent and in the beans from 0·0009-0·004 percent (Duclaux). Copper in similar amount is found in all kinds of beans and husks, and its presence is due to the absorption of copper by the plant from the soil, whence it gradually accumulates in the fruit.

b) The Cacao Shells.

Most of the constituents which exist in the cacao kernels are also to be found in the husks and the methods for isolating and determining them are the same in both cases. The composition of the husk, according to Laube and Aldendorff[89], is as follows:

Table 15.

Key to Columns

B. Amount of husk
C. Water
D. Nitrogenous substance
E. Fat
F. Non nitrogenous extractive
G. Woody fibre
H. Ash
I. Sand

124

	B	C	D	E	F	G	H	I
	Per cent							
Caracas	20·09	7·74	11·68	5·99	35·29	12·79	8·32	18·62
Guayaquil	—	9·11	12·94	10·75	47·08	13·12	6·79	0·21
Trinidad	14·04	8·30	15·14	4·23	46·05	18·00	7·06	0·92
Puerto Cabello	14·92	6·40	13·75	4·38	47·12	14·83	6·06	7·46
Soconusco	18·58	6·48	19·12	6·48	39·39	15·67	8·15	4·71
Mean	16·33	7·83	14·29	6·38	45·79	14·69	7·12	5·90

Zipperer's analysis[90] of the unroasted husks gave the following results

Table 16.

Key to Columns

B. Surinam
C. Caracas
D. Trinidad
E. Puerto Cabello
F. Machala
G. Port au Prince

	B	C	D	E	F	G	H	I
	Per cent							
Moisture	13·02	11·90	13·09	12·04	—	—	—	12·51
Fat	4·17	4·15	4·74	4·00	—	—	—	4·23
Cacao tannic acid soluble in 80% alcohol	5·10	3·80	4·87	9·15	—	—	—	4·58
Theobromine	0·33	0·30	0·40	0·32	—	—	—	0·33
Ash	7·31	16·73	7·78	8·99	—	—	—	10·20

125

Woody fibre	14·85	17·99	18·04	15·98	—	—	—	16·71
Nitrogen	—	2·25	2·13	—	—	—	—	2·19
Proportion of husk in the raw seeds	14·60	15·00	14·68	12·28	16·14	16·00	18·68	15·34

Roasted cacao husks contain according to G. Paris[91] the following constituents:

Moisture 12·57 percent, nitrogenous substance 14·69 percent, fat 3·3 percent, extractives 45·76 percent, crude fibre 16·33 percent and ash 7·35 percent.

50 grammes of the husks when boiled with 500 grammes of water give 25·08 percent extract, 20·68 % organic substance, 4·4 % ash, 0·21 % sugar (reducing substance), 0·79 % theobromine, 0·12 % percent acid, calculated as tartaric acid.

The following constituents have been found by R. Bensemann[92] in the ash of cacao husks:

T a b l e 17[93].

	Maracaibo	Caracas	Trinidad	Machala Guayaquil	Porti Plati
			Per cent		
Ash dried at 100° C.					
I. insoluble in dilute hydrochloric or nitric acid:					
a) Volatile dessicated at 100° C.	0·113	0·421	0·979	0·306	1·24
b) Fixed at red heat	1·917	47·711	29·315	37·662	51·51
II. Soluble in dilute hydrochloric or nitric acid:					
c) Potassium oxide K_2O	31·517	11·812	25·866	23·117	12·17
d) Sodium oxide Na_2O	4·188	3·298	2·726	1·210	2·78
e) Calcium oxide CaO	10·134	4·458	5·097	3·503	4·40
f) Magnesium oxide MgO	9·546	4·703	5·206	4·837	4·09
g) Ferric oxide Fe_2O_3	0·647	0·931	0·339	0·958	0·46
h)					

Aluminium oxide Al_2O_3	0·281	1·554	0·710	1·854	1·04
i) Silicic acid SiO_2	1·180	7·975	2·416	4·321	6·78
k) Phosphoric anhydride P_2O_5	9·068	7·630	4·703	7·288	7·24
l) Sulphuric anhydride SO_3	3·041	1·478	3·398	1·741	2·01
m) Chlorine Cl	1·005	0·220	1·022	0·255	0·44
n) Carbonic anhydride CO_2	25·454	5·399	16·290	11·834	4·24
o) Water H_2O	2·135	2·499	2·263	1·171	1·66
p) Oxygen O equivalent to chlorine	0·226	0·049	0·290	0·057	0·10

As evidenced in the preceding examples, data as to the constituents of the cacao husk deviate considerably with different authors. Laube and Aldendorff, for instance, found 14-20 percent, while Zipperer obtained 12-18 percent of husks.

These discrepancies are mainly due to adhering sand and ferruginous earth collected during the drying and fermenting processes. If the beans are carefully collected and kept free from earthy substances, the percentage of husks as against that of the bean will appear much lower; it is, indeed, now possible to obtain properly treated beans which contain on an average only some 10 percent of husks, such as Ariba and Machala. The husks of these two varieties are exceedingly woody, and their amount sometimes reaches 15 per cent. The latest machinery for cleaning the beans effects so complete a separation of the husks from the kernel that very little of the former remains in the finished cacao preparation (less than 1 percent in thin-shelled beans and no more than 2 percent in thick-shelled beans such as Ariba). For some years it was not possible to effect so thorough a removal of the husk, so that there was always found an appreciably large amount of shells in the finished preparations, which rendered it difficult to detect adulteration. As, however, the quantity of ash present in the husk is double that in the kernel, it was possible to form an opinion as to the intentional admixture of shells from the increase of ash in cacao preparations. Hence the ash was always required to be determined when adulteration was suspected. Under existing conditions the addition of a quantity of shells sufficient to increase the percentage of ash present in the powder or chocolate is scarcely practicable, so that, for the purpose of detecting small additions, other methods must be resorted to, such as the estimation of the crude fibre or silica in the ash[94] with the aid of the microscope, in which it is possible to easily distinguish the forms of the cotyledon (kernel) mass and those of the husk. The diagram on page 14, Fig. 3, clearly shows the elementary forms of the cacao husk as represented by Mitscherlich. It illustrates a longitudinal section of the husk of Bahia beans, enlarged about 500 times, with six different cell elements in alphabetical order. First the compressed cells of the epidermis are to be seen on the exterior, in several parallel series and succeeded by moderately broad and thin-walled cellular tissue of the parenchyma, which sometimes presents large empty spaces (sch) the results of the loosening of the cell walls through the formation of mucilage. This cellular tissue (lp) is also permeated by bundles of spiral vessels (gfb), which, with the dry cells, are characteristic of the husk, as they exist only in very small quantity in the kernel. Then follow parallel rows of cells (lp) resembling epithelial cells; next comes a layer of cells with thick walls, the dry cells (st) and finally several rows of elongated ones (lp). The silver membrane (is) interposes between the husk and the kernel, fragments of which remain adhering to the shell after separation of the latter.

To conclude, we find that the husk of the cacao bean consists of the inner coat of fruit, called endocarp and other parts of the fruit covering, as well as the skin of the seed[95]. The following layers may be distinguished;

1. The pulp, (f in fig. 3) fragile large cells with frequent hiatus;

2. the e n d o c a r p (fe), a single layer of fragile, very narrow and irregularly arranged cells, but w i t h o u t h i a t u ş

3. the e p i c a r p, or skin (se), polygonal and extended cells, with an outer wall of some thickness.

4. the p a r e n c h y m a or cellular tissue (lp), consisting of large and multiform cells, with vascular bundles (gfb), the large mucilagenous or slime cells (sch) and

5. the s k l e r o g e n o u s or d r y c e l l s (st), a single layer of vessels shaped like a horseshoe, and thickening towards the interior, and in conclusion

6. the s i l v e r m e m b r a n e (is), belonging to the earlier inner coat of fruit, and consisting of two single rows of fat-bearing cells.

In examination of the husks of the plane surface enlarged 160 times (fig. 8), it will be noticed that the characteristic epidermis (ep) consists of large and rather elongated but irregular polygonal cells. Frequently on the epidermis may be remarked a delicate network of the cells constituting the fruit pulp (p). Beneath the epidermis lies a very delicate transverse cellular layer (qu) followed by the parenchyma, as already stated. The remaining elementary forms are not readily observed on a plane surface but only in section, though we adjoin a few diagrams, showing the layers as isolated from the pericarp; namely, fig. 9 parenchyma, a layer of sklerogenous cells, fig. 10, and the silver membrane (is) with two superjacent Mitscherlich particles (tr) in fig. 11.

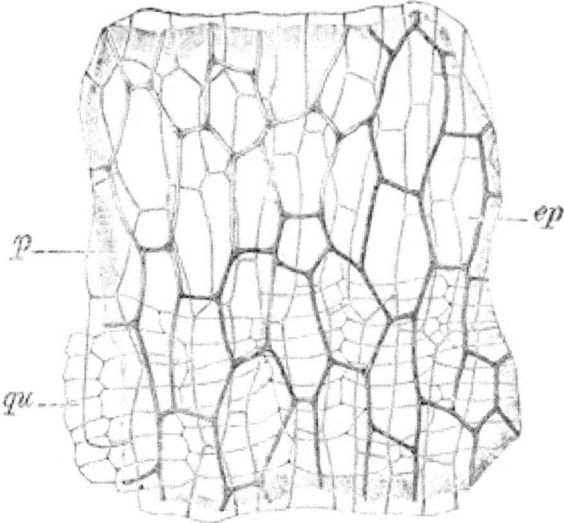

Fig. 8.

131

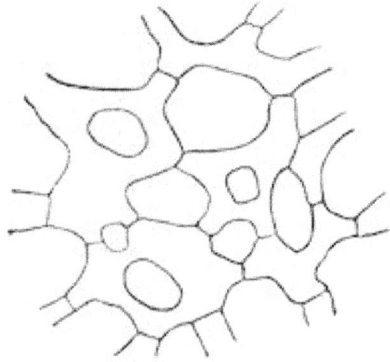

Fig. 9.

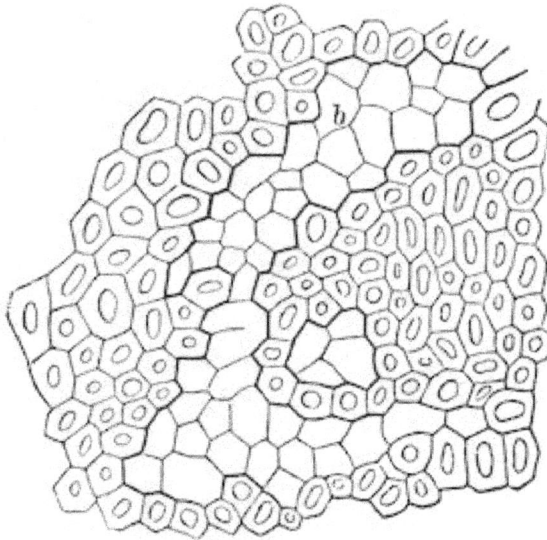

Fig. 10.

For microscopical examination, the husk must first be defatted with petroleum or ordinary ether and then treated with dilute chloral hydrate (8: 5) to assist the definition of the forms. An approximate estimation of the amount of husk in a cacao preparation can be made by means of the microscope, adopting Filsinger's[96] levigation method, which

132

consists of concentrating those elements of the cacao which are seldom seen even in suspension in water, and which sink to the bottom when repeatedly stirred in that liquid. To these belongs first of all the husk, and its presence and determination in the levigation method is accordingly greatly facilitated. The details of the method will be further described in treating of husk admixtures in cacao preparations.

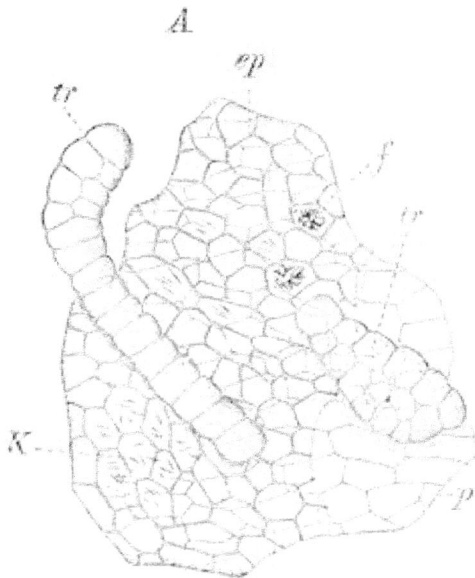

Fig. 11.

Cacao shells are the only by-product in the cacao industry, and have been developed and exploited to such an extent, that a rational utilisation of the ever increasing quantities has become a matter of urgent necessity. They are not used in our industry, for an admixture of husk is not permissible, even in the inferior kinds of chocolate or cocoa powder, but must be regarded as an adulteration. It is true that they have been brought on the market as cocoa tea, and again, have been coated with sugar, to make them tasty;

and to this day, candied husks constitute a favourite sweetmeat of the population of East Germany. But in this way only comparatively inferior quantities of the by-product were absorbed, and consequently projects of all kinds have been suggested to use up larger percentage. As we have seen, the fatty contents of the bean can be extracted with benzine, and there is a resultant 4 or 5 percentage of fat of inferior value, which is commercially known as "Dutch IIa Cacao Butter"; the defatted shells can be further used for the preparation of theobromine, as Zipperer has already noted in the first edition of this book.

Kathreiner's successors in Munich[97] employ an extract of cacao shells prepared with hot water, in order to improve coffee berries during the roasting and to give a flavour to the coffee substitutes prepared from corn and malt. Cacao extract is also prepared from the shells[98] by first treating them with water or steam, and afterwards extracting with water, and finally evaporating as far as necessary. The thick extract thus prepared contains theobromine, and is intended for use either alone or as an addition to cacao powder and chocolate.

Strohschein in Berlin[99] prepares from the shells a thick liquid extract which he calls "Martol Its preparation was suggested by the fact that the cacao husk gives evidence of containing a considerable amount of iron. In "Martol", the iron occurs as a tannate, and the preparation further contains theobromine, carbohydrates, and phosphoric acid. The preparation is said to be used as a medicinal remedy in chlorosis, yet has scarcely justified such a statement.

Alfred Michel of Eilenberg[100] utilises the shells in the preparation of a brown colouring material. The husks, free from impurities, are first soaked in soft water, with or without the addition of sulphuric acid, then washed and finally treated with a strong 35 % solution of caustic soda. From the alkaline solution, the colouring matter is precipitated with acid or acid metallic salt, collected on a filter, and again washed. Thus obtained, it is a dark reddish-brown paste, possessed of a vitreous fracture. The yield of colouring matter is from 20-25 % of the weight of the original shells. By re-treatment with alkali, the paste can be again obtained in solution and can be used as required, either in liquid or paste form. The colouring matter can be obtained in different tints, either by soaking the shells in more er less dilute sulphuric acid, or by precipitation from the alkaline solution at various temperatures, or yet again, by the addition of metallic oxides.

Boussignault[101] says that in Paris briquettes have been made from cacao shells, and twenty-two years ago, Zipperer[102] proposed to use them as fodder, especially for horses. Experimental work in that direction was instituted, but for various reasons, had to be abandoned. The question as to a rational working up of the husk of the cacao bean is once more receiving special consideration, more particularly since the publication by the "Association of German Chocolate Manufacturers" of a prize essay on the subject. The fodder value of the husks as determined by Märcker is apparent from the following figures:

Table 18.

Shells	free from dust, whole %	fine meal %	whole and dusty %
Moisture	9·08	6·50	9·95
Albumen	13·56	14·13	12·69
Albumen digestible	6·06	7·07	4·38
Fat	2·65	6·76	3·96

Raw fibre	29·14	25·80	21·55
Ash	6·32	6·44	7·26
Non-nitrogenous extractive	39·25	40·37	44·59

Feeding experiments which were carried out in certain agricultural institutes showed that the cacao husk stands in nutritive value between good meadow hay and wheaten bran, and is not only a fattening fodder for oxen, but also a valuable feeding material for cows and deer[103]. These results have been confirmed by Prof. Feruccio Faelli in Turin[104].

The advantages of cacao shells as fodder, when a comparison with bran is established, are at once apparent. Two hundredweight (that is to say, about 220 lbs. averdupois) cost only from six to seven shillings, whilst the price of bran varies between nine and ten shillings. The husks also keep better, for after having been stored eighteen months, Professor Faelli found that they had undergone no alteration, whilst on the other hand bran had become sour. A further advantage possessed by the husk is that it will absorb four times its weight of water against three times absorbed by bran. Cattle not only readily get accustomed to the fodder but subsequently take to it with eagerness. The best results were obtained with Dutch, Swiss and Parmesan milch cows. After 10 days feeding the butter and milk-sugar had increased, as well as the daily average yield of milk from 44 to 49·5 kilogrammes. As soon as the feeding with cacao husk was discontinued the yield of milk decreased. Faelli concludes that cacao husk, which can be used as a fodder up to 4 kilog. daily, exercises a very favourable influence on milch cows, and he purposes to continue the investigation with horses.

In a report on the Experimental Farms of Canada 1898,

page 151, reference is made to the manurial value of the husks in enriching the soil with nitrogen and potash, a fact which had already been pointed out by Boussignault.

The future use of the husks appears therefore to be ensured, and it is to be hoped that it will allow of a permanent consumption of this by-product.

Part II.
The Manufacture of Cacao Preparations.

A. Manufacture of Chocolate.

The Preparation of the Cacao Beans.

Up to the end of the eighteenth century the manufacture of chocolate was carried on entirely by hand, a method at once laborious and inefficient. The workman used to kneel down on the ground, and crush the beans in iron mortars. It was not until 1732[105] that Buisson introduced the use of a bench and so rendered that inconvenient and unwholesome practice unnecessary. Even to-day, the Chinese cooks on the Philippine islands carry their chocolate "Factory" about with them, in which the trestle is essential. It further comprises a small marble mortar and warmed pestle, and by means of these utensils and implements the hulled beans are pounded, and the triturated mass so obtained spread out. It is then flavoured with sugar and spices. With that exception, hand labour in the chocolate manufacture has since the year 1778 been entirely displaced by machinery, when Doret exhibited the first specimen before the medical faculty of Paris. According to Belfort de la Roque,[106] a Genoese named Bozelly had already constructed a mill by means of which he was able to prepare from six to seven hundred pounds of chocolate daily, comparing favourably with the thirty pound output yielded by hand labour. Pelletier[107], in 1819, describes a machine for the mechanical preparation of chocolate of his own construction, capable of doing the work of seven men. The machines used in the chocolate manufacture have since that time been repeatedly improved and re-constructed, although always with this

138

one end in view, namely to obtain a fine even cacao mass, and afterwards mix it as thoroughly as possible with the other ingredients employed.

The first machines of the modern type were constructed by the Parisian mechanic George Hermann (1801-1883) in the year 1830, to which inventor we are indebted for the principle of fine grinding with varying velocities, on which manufacture of chocolate is based to-day. There is at the present time a rather large circle of manufacturers engaged in the putting together of special machines for the preparation of cacao and cacao products, chocolate apart.

Whether chocolate manufacture be carried out on a large or small scale, it always involves the subjecting of the cacao bean to a regularly succeeding series of operations, before the resulting product known as "Chocolate" (in the strict commercial sense of the term) can be obtained.

The respective operations succeed each other as follows:

I. Preparation of the Beans.

1. Storing, cleansing and sorting of raw beans.

2. Roasting the cleansed beans.

3. Crushing, shelling and cleansing the roasted bean (removing the radicles etc.)

4. Mixing different kinds of beans.

II. Production of the Cacao Mass.

5. Grinding the beans till they yield a homogenous paste on heating.

6. Mixture of the licuefied cacao mass with sugar,

spices, etc.

7. T r i t u r a t i o n by rollers.

III. P r e p a r a t i o n of the resulting C h o c o l a t e.

8. E x t r a c t i o n of air, division and m o u l d i n g.

9. C o o l i n g.

10. P a c k i n g and s t o r i n g.

This represents the general course of manufacture, which we will now proceed to describe in more detail, following the headings given above.

---

1. Preparation of the Beans.

1. S t o r i n g, c l e a n s i n g and s o r t i n g.

Right up to the moment when they are to be used in the manufacture, the raw cacao beans must be kept as originally packed, and stored in an airy sun-lit room; although if they have accumulated moisture during transport or sustained any manner of damage in harvesting, they should then be emptied out of the sacks, spread out over the floor of such a room as above described, and dried as effectively as possible. It has also been recommended that such beans be washed with a dilute solution of caustic potash (1 in 5000), and afterwards dried rapidly.

Unfermented beans, those damaged in the harvest, and those which have received no proper fermentation, develop a greyish white colour with occasional tints of violet and an unpleasant, bitter herbal flavour, properties which unfortunately penetrate to the resulting cacao products. Attempts have been made to meet this evil with a so-called "Secondary Fermenting Gordian[108] proposes in this connection that the beans be filled in water-butts, and steeped in warm water for at least 48 hours (so that obviously the butts must be kept in a warm room), at the expiration of which time it can be poured off, and the beans dried in a chamber heated to a temperature of between forty and fifty degrees centigrade. There is said to ensue an appreciable improvement as to flavour and colour, when this process is carried out.

The magazines in which cacao beans are stored have sometimes an unwelcome visitor, to wit, a grub which according to W. Hauswaldt[109] happens to attack just the best kinds of Caracas and Trinidad. As eggs of the grub have on several occasions been found on the interior of the still unshelled bean, we may assume that they were deposited by a butterfly (species unknown, but possibly Ephestia cahiriteller, cf. von Faber loc. cit. page 335) either before or immediately after fermentation, and no later. Sometimes these grubs appear on the surface of the sacks, which they overspread in a few days. Removal of the infected packages, opening the sacks, and exposure to the sun, as well as a thorough cleansing of the storehouses, is attended with a qualified amount of success. The best plan is to destroy the moths during their period of activity in the summer months June, July, and August.

According to Hauswaldt, Stollwerck[110] and G. Reinhardt[111], this can be effected by placing in the store rooms large, shallow basins of water, near which burning petroleum lamps are introduced on the approach of dusk, favourably placed on a pile of bricks and stone, so that they

141

clearly illuminate the reflecting water. The moths assemble round the light en masse and either perish in the water or flame, a fate which sometimes overtakes even the larvae, for they display the same fatal attraction for any light, real or apparent. The water must be changed every day, as otherwise the wing-dust collecting on its surface affords a means of escape to the insects coming later. As the weather becomes cooler, the doors and windows of the store-rooms should be left open, so that when frost sets in, the rest of the maggots may be destroyed.

The cleansing and sorting of the raw cacao bean is the most important factor in the manufacture of chocolate, and yield a manifold return, for inferior and cheaper kinds of bean which have passed through these processes can be advantageously mixed with finer varieties. The chief object of cleansing and sorting is the removal of foreign bodies and such chance admixtures as sand, pebbles, and fragments of sacking, which are liable to damage the stones used in grinding at a later stage of the preparation, or communicate an unnatural and disagreeable smell to the subsequent roast products. These admixtures are so multiform and various that they cannot be removed solely by the aid of machinery, but must be finally picked out by hand. Mechanical appliances are limited to the removal of pebbles, dust, and possible fragments of iron, after which preliminary cleaning the beans are thrown on straps, where they can be picked by hand. The collector of these foreign bodies would find himself with a rather interesting stock at the end of a few years, as Wilhelm Schütte-Felsche points out.

The cleansing of the raw beans was formerly carried out in so-called roller casks, placed horizontally, and revolving round an axle fitted in the floor, whence it passed upward, cutting them slantwise. In this apparatus the beans were rolled and vigorously rubbed together, and afterwards the hand-picking succeeded. More recently, the roller casks have been displaced by rotary cylindrical sieves, driven by motor power.

Such a machine is illustrated in fig. 12. The beans are lifted to a rotatory cylindrical sieve by means of an elevator, where they are freed from dust and dirt; in other sections of the sieve fragments of blossom, sacking, or cloth are isolated, whilst occasional splinters of iron are removed by a large magnet. So prepared, the beans are cast on running belts, and here the hand-picking above-mentioned is carried out.

Fig. 13 shows a cleansing machine for the same purpose, which has recently become rather popular. Here the dust passing from the sieve is sucked up into a dust chamber, by means of an exhauster, whilst pebbles, blossom fragments, and small beans are separately isolated. The cleansed beans pass likewise under magnetic influence, which removes traces of iron, and finally succeed to the running belting.

Often the beans are introduced into an extensive brushing machine before roasting, to cleanse them from dirt etc. These are generally found in such factories as have circular and cylinder roasters with direct heating apparatus. Fig. 13 a shows such a brushing machine for cacao beans.

2. Roasting the Beans.

The cleansed and sorted beans are now subjected to a high temperature, that is to say, they are now roasted. This roasting answers many purposes;

1. The aroma and flavour of the bean is so developed.

2. The starch granules are gelatinised.

3. The herbal constituents are so transformed that the flavour of the beans becomes milder; a distinct improvement.

4. In the consequent drying, the shells are rendered brittle, and more easily removeable.

5. The beans themselves can afterwards be better ground.

The roasting of the cacao bean does not demand so high a temperature as that of coffee, to effect the above chemical and physical changes. Experience has shown that the best temperature lies between 130-140 ° C., though deviations from this standard have recently become frequent and considerable, according to the uses for which the cacaos are intended, and roasting has sometimes taken place at a temperature even as low as 100 ° C.

Fig. 12.

The process of roasting can be carried out in the roasting drum or machine in a variety of ways, as:

1. Direct roasting over a coal fire,

2. Passing of a hot-air stream over the beans,

Fig. 13.

3. Roasting by means of gas, with compressed air, as far as s o u r c e s o f h e a t are concerned; and as regards s h a p e o f t h e d r u m, it is to be noted that the cylindrical are most in use. The separation of the shells from the kernel was still effected at the beginning of the present century by stirring the beans in water and so detaching the inner coating of the seeds, the method adopted by Weisched (Mitscherlich page 112). Not till this stage had been reached were they subjected to a strong heat, causing the shells to spring off.

This method has at the present time only historical interest, for the so-called roasting drums, as used in the preparation of coffee, are now universal.

Fig. 13a.

Roasting must be attended with the greatest care, in order that it may neither be too thorough nor insufficient. It is a great mistake to think that the roasting machine can be handed over to the care of any apprentice. That nicety of roasting which corresponds to the variety and its subsequent utilisation constitutes the qualitative basis of the chocolate manufactured later. It is impossible for even the best chocolate maker to retrieve what has been spoilt in this important preliminary operation, wherefore a skilled workman, endowed with a keen sense of taste and smell, is always to be seen at the roasting machine.

It has already been attempted to provide a means of security against over-burning by the construction of the so-called safety-roaster, about which will be spoken later.

Overroasting is immediately indicated by a disagreeable empyreumatic odour (resembling that of roasted coffee); the husks char and the kernels crumble, also betraying a charring on the outside. There is a correspondingly

147

increasing keenness of flavour, and a transference of theobromine from the kernel to the husks (cf. page 65). From the destructive distillation of the cacao fat arises that volatile and pungent acroleine which is the principal cause of the empyreuma of the over-roasted bean.

Fig. 14.

The following general precautions in roasting cacao are worthy of note; 1. the beans should not remain too long in the roasting drum; 2. they should be kept on the stir, for which reason the apparatus is made revolvable on its axles; 3. the heat applied should be carefully regulated; and 4. to guard against a loss of aroma, the roasted beans should be cooled as rapidly as possible.

As the cacao must be more or less roasted according to its quality and ultimate destination, which entails the acquisition of considerable empirical knowledge on the part of the workman entrusted with this process, it would be neither advisable nor practicable to annex definite instructions as to time and temperature requirements.

In the following we describe a machine which is to be found in most factories and which corresponds to all the demands of technique. From its heating system, it belongs to the class of hot-air current roasters—direct coal fire assisting—and in shape to the cylindrical roasters.

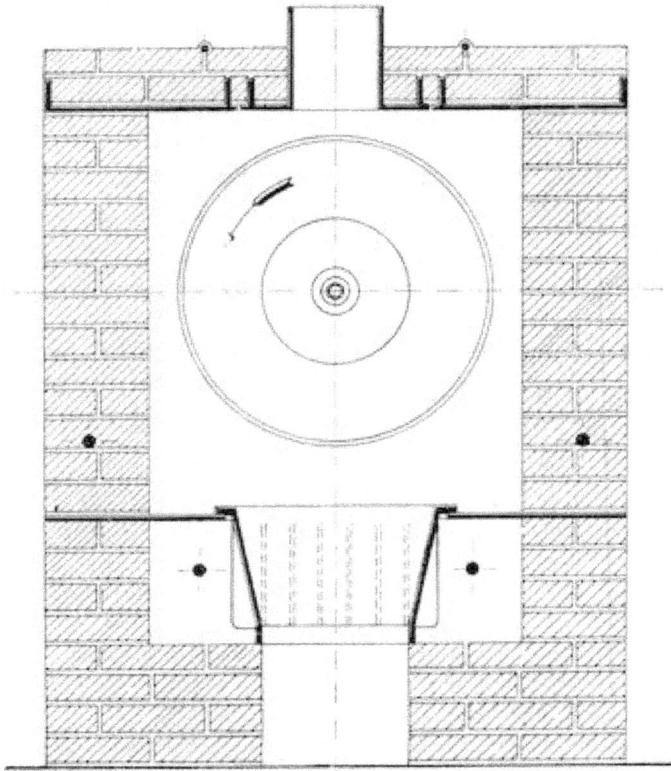

Fig. 14 a.

This machine is illustrated in fig. 14 and shown in section
in figs. 14 a and 14 b. To prevent loss of heat by radiation, to
save fuel, and preclude possibilities of danger from fire, the
whole installation is walled in. Driving shafts occur at the
back of the machine, and the charging apparatus is
introduced in front. A furnace lies directly under the drum,
whilst on either side are chambers accessible to currents of
fresh air, which are provided with heating tubes and which
admit of a regulation of the air supply. They are shut off
from connection with the gases from the fire, so that only
the fresh air heated here can penetrate to the roasting
products in the charged drum. There are winnowing

150

shovels fitted in this, calculated to keep the beans in motion
and facilitate the access of air. When the hopper is closed, the
gases arising from the roast product can be led off by an
annexed outlet pipe, and thereupon condensed and the
resulting liquid drained off at the foot of the machine. For
the attainment of the proper degree of roasting, as well as
for controlling the whole process, there is a sampler to every
machine. The drum is emptied whilst in motion, its door-
like front being turned aside and the roasted beans
transferred by the winnowing shovels before mentioned to
trolleys wheeled underneath.

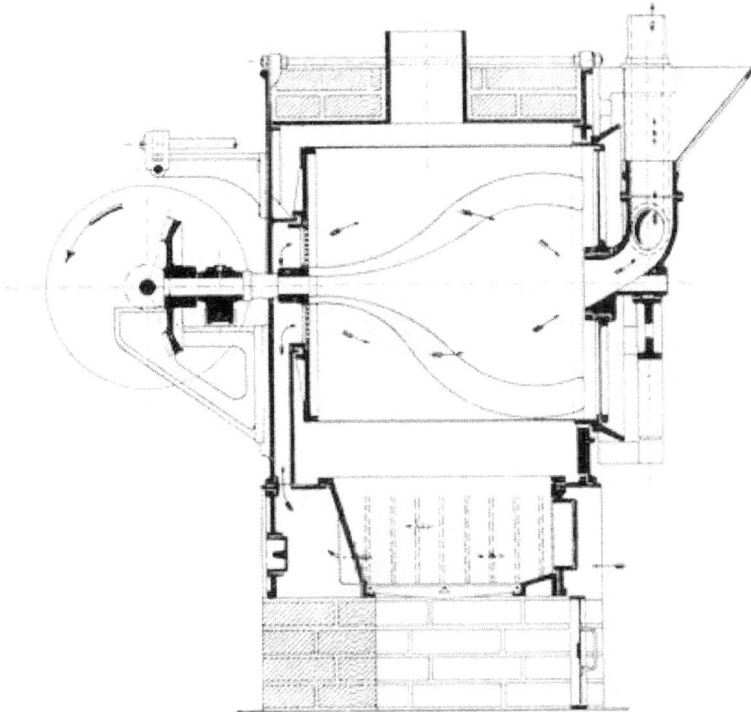

Fig. 14 b.

The loss of heat by radiation is very insignificant, as the
machine is completely walled in. Any kind of fuel may be

used. Since the stoking as well as the removal of soot takes place at the front, several of these roasters can be set up side by side. It is a great advantage of this installation, that by removal of the front of the drum its interior is laid quite open, admitting of a thorough overhauling which is attended with every disadvantage for the flavour of subsequent roasting lots.

The machine here described is constructed in varying sizes, with an outside capacity of four hundred kilograms.

As already mentioned the so-called safety-roaster offers a certain security against the burning of the beans as the roasting boiler is lifted out of the fire by means of an automatically working safety regulator. Figs. 15 a and b show a spherical roaster open and closed.

Fig. 15 a.

The principal of construction is founded on the fact that each roasting is connected with a loss of weight and it is logical that the same quality cf beans always yields the same loss of weight at a certain degree of roasting. On an average cacao yields a loss of 6-7 %. According to this, the loss of weight which can at first be empirically ascertained, for example by a new kind of bean, can be calculated and can be indicated on a regulator, on the principle of the Roman scale. When the beans have lost the weight in question the counterpoise of the regulator raises the axle of the roasting sphere by means of which the working of the whole machine is set in motion.

There is no exception to the rule that only beans of one and the same kind should be roasted and broken up together, as thickness or thinness of the shells determines to a large extent the time required for roasting, and also an even size of bean is necessary to the smooth operation of the breaking machine. The husks of the roasted cacao bean are hygroscopic, and consequently the roasted unshelled beans contain more moisture after having been kept for a time, than they do in the raw state; but the drier the bean is, the easier it shells. The cacao is therefore to be worked up as quickly as possible, or at least kept in well covered metal boxes till further treatment can be proceeded with.

Fig. 15 b.

As sources of heat we find direct and indirect stoking with house coal and coal gas, and besides these, for the installations of larger factories Dowson gas is especially suitable, as it does not involve too high a temperature, and the outlay is not so great as when coal-gas is used.

Fig. 16.

The roasting machine in fig. 16 for Dowson or coal gas belongs to the class of roasters with direct firing. It corresponds to the one diagrammed in Fig. 15 as regards charging and emptying. Here also the front wall of the drum can be removed, and the interior consequently laid completely open. The transmission of gas is effected at an air pressure of one atmosphere, for the attainment of which an air pump is fitted up in the vicinity, capable of feeding four machines at the same time. The drum holds about 150 kilos. It goes without saying that the regulating of the requisite heat is in this instance of the utmost ease and nicety. Another preponderating advantage of this machine as compared with those heated with coke or ordinary coal is its clean operation and the extraordinary speed with which it can be both started and stopped. Form 3-4 cbm. of coal gas are needed for 100 kilos of beans, whilst for Dowson gas, which has not such a high heating value, much larger quantities are required, and consequently a stronger framework becomes necessary, though here no air pumps need be put in operation.

Fig. 17.

Steam roasting apparatus have not proved particularly successful, as has been evident in all experiments hitherto made with them, and steam agency does not appear to be suitable for the cacao bean, it admitting of no thorough and at the same time even roasting.

Yet on the other hand the hot air-current roasters described enjoy an ever increasing popularity, partly because they are heated indirectly, and again because they appreciably diminish the time taken up in the actual process, which in other cases approaches to as much as thirty or forty minutes, without exposing the beans to the danger of burning or getting charred.

As just stated, the beans should be passed on to the next process as speedily as possible, yet on the other hand be completely cooled off, so as to loosen their shells before they arrive in the breaking machine. There are also special constructions for this cooling. If the roasting drums are fitted up directly on the ground, it is effected by disposing the beans issuing from these machines in wide baskets or sieves, and letting them cool there before bringing them to the next process. Should they be situated at a sufficient height, the beans can be slowly transferred down a shoot connected with the rooms below, where crushing mills await them, and cooled on the journey by a play of fresh air currents.

Very much to the purpose and well adapted as regards most of the requisite conditions, are the cooling trucks with exhaust apparatus shown in fig. 17.

These trucks are fitted with perforated false bottoms and with sliding shutters at the side. After the contents of the roasting machine have been discharged into the trucks, these are wheeled over to the exhaust apparatus easily recognisable in the diagram, where the cacao is so far cooled that subsequent "after-roasting" is impossible, whilst the gases given off are conducted by the ventilator. This exhaust chamber can be made to work from both sides.

3. Crushing, hulling and cleansing.

Up to ten years ago, the crushing and shelling of cacao beans had not been so far perfected as to effect the complete separation of husk and radicle from all particles of kernel, or to prevent loss by isolating and collecting the minute particles of kernel, which are drawn up through the exhaust apparatus in conjunction with the lightest of the cacao shells. Yet the requirements demanded of a satisfactory machine advanced to such an extent that not only cacao nibs free from shell were postulated—an end scarcely hard to attain—but shells free from cacao nibs were made a further essential. A machine which performs both these objects not only works excellently, but is also economical. For a solution of this problem the Association of German Chocolate Manufacturers, which is specially interested in all that concerns the chocolate industry, offered a prize years ago; the firm of J. M. Lehmann were the first to construct a machine answering every call made on it to perfection.

Fig. 18 illustrates a crushing and cleansing machine averaging an output of 2500-3000 kilos, of the latest and most modern type.

Fig. 18.

The beans are first broken into smaller pieces in all machines now employed as crushing, shelling or cleansing apparatus, and the one at present under consideration provides no exception. An air-current is made to play on these fragments, which finally isolates and transfers the loosened shells to another part of the apparatus. The cacao next succeeds to a crusher of regular capacity lodged in the upper part of the machine, being despatched on an elevator. The fragments fall into a cylindrical sieve, dust being detached in the first compartment, whilst the meshes of subsequent compartments gradually increase in size and sort the products therein transmitted in corresponding sizes. There is a groove traversed by air-currents—proceeding from a ventilator—immediately under each compartment. This current of air can be regulated, i. e. made weaker for lighter and stronger for heavier fragments, and there is a ventilator for every compartment to make this regulation of the easiest, and in this way shells of equal size but specifically lighter than, the cacao fragments are most efficaciously separated. Contrasting with the older type of machine, it works almost noiselessly, all shakings of grooves and sieves being entirely avoided; in addition to which there is a perfect exclusion of dust, when the shells are transferred into the dust-removing chamber. A further advantage is that there is no wearing out of the machine, except as regards the direct crushing apparatus, which occasionally need renewing.

The dust particles before mentioned, which possibly comprise as much as one half of the cacao fragments, require a special kind of working up, on d i f f e r e n t m a c h i n e s, before the cacao still contained therein can be obtained. It is a fact obvious and apparent, that the smaller the fragments of shell mixed with this crushed cacao, the more difficult will be their separation, a fact of equal importance to

159

technical and analytical science, and the more scrupulously this process is to be carried out, the greater the lavishment on sieves and ventilating compartments entailed.

To effect this operation on the breaking machine is seriously to overtask the latter, and defeats its own end, as experiments carried out in the Chocolate factory of Schütte-Felsche have proved, inasmuch as it leads very easily to mixing of the products which are to be kept separate.

Fig. 19 shows such a machine for the cleansing of this so-called cacao "dust

The particles are raised to a large flat sieve by means of an elevator, again sorted in different sizes, and submitted to air currents of corresponding strength. The quantity obtained varies according to the variety of cacao, though in some cases it may amount to 50 or 54 percent. What remains after this process is absolutely worthless and can only be considered as refuse, at least as far as the chocolate manufacturer is concerned.

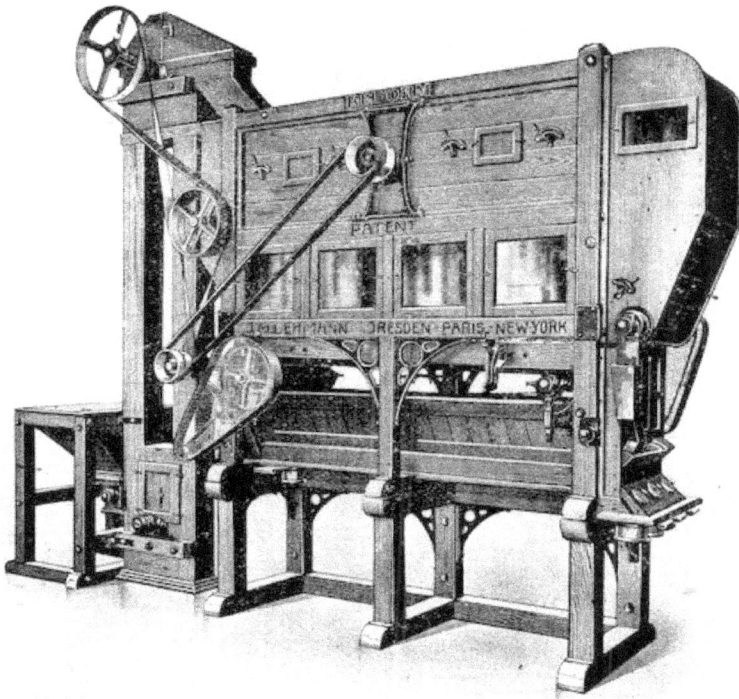

Fig. 19.

It has become necessary in modern manufacture that iron fragments occurring in the machine not only be separated by distinct magnetic fields in the respective machines, but that this also be effected in a machine specially constructed for the purpose. Fig. 20 illustrates such an electromagnetic apparatus. The advantages of this system are that it avoids magnets limited in strength, and by the functioning of strong electro-magnets perfect cleansing even in the case of the largest output, as well as machines of the most simple construction, can be guaranteed.

We submit the following description of the machine and its method of working.

Fig. 20.

The machine contains a hopper with sloping groove to obtain an even introduction of the beans to be cleansed. At the end of this there is an electro-magnet roller, consisting of a non-magnetised mantle and a magnetic compartment round which it turns.

After traversing the sloping groove, the beans succeed to the roller, meeting it at a tangent. As soon as they reach the field of magnetism, all iron fragments are appropriated by the revolving mantle, whilst the beans themselves do not come into contact with this, but pass directly underneath.

The iron fragments are disposed of separately, and outside the magnetising area.

It is of prime importance in the preparation of chocolate and more particularly of cocoa powder (easily soluble cacao), that the crushed material proceeding from the crushing machine should undergo a further purification, with a view to separating, and removing the hard radicles. These constitute the gritty sediment of insufficiently prepared cacao powder, when dissolved. J. M. Lehmann effects the complete removal of the radicle by means of his machine D. R. G. M. No. 24,989 (Fig. 21).

Fig. 21.

Here the finer siftings from the crusher are transferred to

163

the controlling feeder, under which a small ventilator occurs, which provides for the removal of any still remaining portions of husk. Cacao and radicle descend to a shaking sieve, the finer particles passing through its meshes, whilst the larger grains fall into a pocket attached to the end, as cleansed product. The former fragments now succeed to a cylinder, having its inner surface punched with small cavities (fig. 22) and while the cacao particles remain in those cavities during the rotation of the cylinder, the radicles of more elongated form are caught up by a special separator (1) and so prevented from being carried round with the rest. The cacao particles are then made to fall into a trough (3) by a brush (2) working against the cylinder, and subsequently urged forward by a conveyor (4). That process is enacted all along the cylinder, so that finally cacao and radicle issue from the machine completely separated.

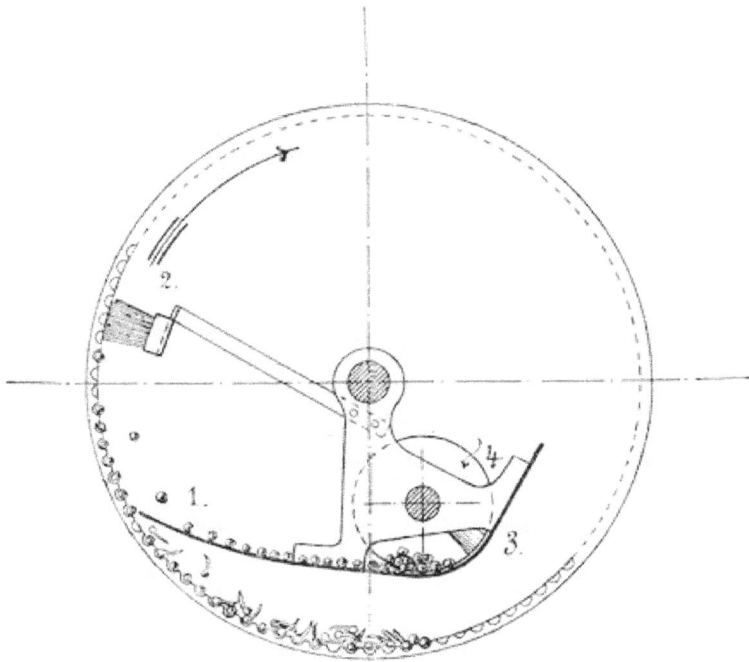

Fig. 22.

The advantages, economical and otherwise, attending the use of the above breaking and cleansing machines become apparent when the following figures, registering results obtained in several experiments, are considered. Formerly the loss experienced in sorting, roasting, crushing and hulling averaged about 30 % of the total beans, but now the employment of the above machines shows the following satisfactory improvements.

The loss of 823 kg Machala beans, unroasted, amounted to a total:

        a) in picking   3·6 kg

        b) " roasting   63·5 "

c) " shelling	61	"
d) " dust	34	"

162·1 kg or 20%,

without taking into account the application of the waste; 2267 kg of St. Thomé raw cacao lost:

a) in picking	5 kg	
b) " roasting	170	"
c) " shelling	152	"
d) " dust	79	"

406 kg or 20%.

According to these data the use of these machines admits of a saving of about 10 percent more material than in former work.

In connection with these particulars it is also of interest to consider the qualitative and quantitative composition of the various waste products of the manufacture. Filsinger[112] has at the instance of the Association of German Chocolate manufacturers, examined a mixture of 50 pounds of large Machala beans with an equal quantity of small beans, after passing it through a shelling machine of the most modern construction, and he thus obtained:

70 pounds of large kernels,
9·2 " " medium kernels,
0·8 " " radicles,
10 " " husk (outer woody shell),
4 " " cacao waste,
6 " " other loss,

The 4 pounds of cacao waste yielded by further sifting:

a) kernel I.   sort 250 grammes,

        II.   "   50   "

       III. "   220   "

       IV.   "   25   "

b) husk I.   "   185   "

        II.   "   55   "

      III. "   370   "

      IV.   "   80   "

c) cacao dust   725   "

d) waste   30   "

e) loss   10   "
_____

        2000 grammes.

Chemical analysis of these portions gave the following results:

	Percentages			
	Ash	Sand	Fat	Fibre[113]
1. Husk 10% of the raw cacao	11·15	1·90	4·50	21·36
2. Cacao waste 4% of the raw cacao	4·80	0·35	15·40	16·31
3. Seed shells I. sort ;0·37% of the raw cacao	6·70	—	21·64	10·29
4. Seed shells II. sort 0·11% of the raw cacao	7·10	—	18·39	8·75
5. Seed shells III. sort 0·74% of the raw cacao	7·20	—	15·76	12·16
6. Seed shells IV. sort 0·16% of the raw cacao	7·80	—	16·40	12·74
7. Cacao dust 1·45% of the raw cacao	11·75	—	22·06	8·40

8. Waste 0·06% of the raw cacao | 7·05 | — | 20·44 | 9·81 |

From these data it is evident that there is a great difference between the chemical composition of the so called cacao waste and that of the exterior ligneous shells. From the large amount of fat present in the former material it might be regarded, in the full sense of the term, as a cacao constituent and, for that reason, its presence in cacao preparations should not be objected to, while the husk containing as much as 20 percent of woody fibre cannot be considered a cacao constituent in the same sense.

4. Mixing different kinds.

Stress has already been laid on the variations in taste incidental to different species of bean. It has further to be noted that they develop a milder and more aromatic flavour according as they have been more properly fermented, and in contrary instances possess an astringent and even acid taste. It therefore becomes an aim of the manufacturer so to improve the flavour of inferior varieties by mixing with the finer as to produce a resultant cacao giving perfect satisfaction to every taste. Nevertheless the general rule still holds good that for the preparation of the finest qualities of chocolate only the better sorts of bean (as Caracas, Ariba, Puerto Cabello etc.) should be employed. For inferior and less expensive ware other varieties of bean suffice, the mixture being obviously regulated by the prevailing market prices.

In many instances the proportions of such mixtures are kept secret by the manufacturer as matters of importance, and every individual manufacturer has his own method and specialities as regards such blends.

We compare here a few verified blends:

	1.			2.

Caracas		Caracas	= 1 part
Guayaquil	} of each 1 part	Bahia	= 5 parts

1.

Caracas ⎫
Guayaquil ⎭ of each 1 part

2.

Caracas = 1 part
Bahia = 5 parts

3.

Maracaibo ⎫
Maragnon ⎭ each 1 part

4.

Trinidad ⎫
Maragnon ⎭ equal parts

5.

Caracas = 1 part
Maragnon = 2 parts

6.

1 part Ariba
1 part Surinam
1 part Trinidad

7.

1 part Ariba
1 part Trinidad
1 part Surinam
1 part Caracas

8.

3 parts Ariba
1 part Trinidad
1 part Surinam
1 part Caracas

9.

1 part Machala
1 part St. Thomas

Ceylon cacaos are not used so much as mixing varieties, but almost exclusively as covering agents, to make other cacaos lighter coloured (sometimes almost approaching yellow).

The beans are weighed off in these proportions on a sensitive scale, and then passed on to be ground and triturated into cacao paste.

II. Production of the Cacao Mass.

5. Fine grinding and trituration.

Formerly the roasted, crushed, and decorticated beans were frequently ground before being transferred to the "Melangeur",—a machine that will be described later—, in which they were then reduced to a finer state of sub-division and lastly mixed with sugar. For this grinding, mills of various construction were employed (as Weldon, Pintus etc.). But as time rolled on the Melangeur took the place of these preliminary grinding mills, and in this it was endeavoured to effect that fine division of the cacao mass which is essential to the production of a homogeneous cacao and sugar intermixture, but without complete success. Cylinder rolling machines (French method) were the first to attain this result.

At the present time, the roasted and cleansed kernels are ground so fine as to become a semi-liquid when subjected to heat, and that is done whatever the ultimate destiny of the cacao, whether it be intended for chocolate or cocoa powder. This object is obtained by means of special mills, constructed with "Over-runners

Fig. 23.

Fig. 24 a.

These cacao mills, which were formerly but seldom met with in chocolate factories, have now become indispensable necessaries, since they have the advantage:

1. of rendering the cacao mass in this semi-glucose form more easily miscible with sugar, a factor of the highest importance for the commoner and cheaper qualities of chocolate;

2. of grinding the cacao as fine as possible in one operation and the simplest manner.

Fig. 24 b.

173

Fig. 24 b.

Fig. 24 c.

But side by side with the appreciation which these mills met with, there arose a corresponding increase in the demands made on them, such as the utmost nicety, greatest possible output, and least possible necessity of after-heating, and these have been successively answered by twin, triple and at the present time even quadruple mills. fig. 23 shows a simple grinding mill which can only come into consideration in connection with the smallest of branches, whilst Fig. 24 a and b illustrates another with three successive stones arranged one above the other, such as will be found in all the larger factories of to-day. Also a triple mill but with grindstones of increasing size pictured in fig. 24 c. A mill possessing four pairs of grinding stones is given in fig. 25, and is calculated to meet each and every conceivable demand.

Whilst simple, double and triple mills are brought on the market in different sizes, corresponding to the outputs required, these quadruple mills are only constructed in the largest sizes. They grind perfectly, and without detriment to the flavour, deliver quantities of cacao figuring at from 1000 to 1200 kilos daily. There is naturally a larger output if the fatty contents of the cacao are considerable, a thorough roasting being always presupposed.

The axles occurring on these quadruple grinding mills are connected with one another by means of spur-wheels, and the axles themselves run in ball-bearings, which not only permits a perfectly noiseless operation of the machine, but also makes the action very easy, that is to say, dependent on only very little motor power. The cacao is raised to the hopper by means of an elevator, where the quantity introduced into the machine is regulated, and then passes between crushers occurring in the middle of the first pair of grinding stones, which it subsequently leaves as a pasty mass. It is then conducted along a groove into the second mill, and here undergoes further grinding, and so to the third and fourth, where the process can be described as trituration, for the cacao leaves the machine in liquid form. Only in this manner is it possible to obtain the finest ground product, without any disastrous accompaniment of excessive heating.

Cacao mills with one stone suffice for the production of chocolate mass on a small scale, but for the manufacture of cocoa powder, twin or triple grinders must be employed.

All these are of the "Over-runner" type, act by their own weight, and consequently do not involve the disastrous consequences which were entailed by the "Under-runners" tried formerly.

About the middle of the nineties of the last century, experiments were made with a view to superseding these types with mills having stones of varying sizes, and first larger upper stones of a grinding pair were tried, then larger under stones, but neither have been able to maintain themselves in the workshop, and the grinders of equal size still hold good as the fittest and most popular.

Fig. 25.

Attempts have recently been made to introduce a machine combining mill and roller. Its value lies in the fact that with a relative increase in the grinding rapidity, it does not involve a greater than requisite heat, and on emerging from the machine the cacao shows no deficiencies as to flavour, and is withal much finer than that produced in other processes.

Fig. 26.

Fig. 26 shows such a machine. The mill on this serves merely to reduce the hard kernel to a pulp, and this admits of the grinding stones being placed farther apart, and so occasions no heat. Trituration is then effected by a roller apparatus, for which operation machines with four rollers have been proved most satisfactory. As such roller machines are furnished with water-cooling systems, it is possible for the cacao to be kept cool even on these.

6. Mixture with sugar and spices.

Fig. 27.

A thorough mixing with sugar can only be effected when the cacao paste is heated to a temperature rather above the melting point of cacao butter, that is to say, as high as from 35° to 40° C., and consequently the incorporating machine in which that operation is carried on is provided with a steam jacket. For this process it is advisable to have the chocolate in a semi-liquid condition, wherefore the ground cacao issuing from the mills is transferred to steam-heated vessels (fig. 27) fitted with taps suitable for drawing off the mass as it is required. Formerly the cacao mass was fed into the melangeur in lumps and there liquefied. But as this necessitated the application of heat to the melangeur, attended with the risk of cracking its under-plating, and also a postponement of the mixing processes, whereby considerable time was lost, this method no longer obtains

179

to-day. It is at present usual not only to warm the cacao mass beforehand, but the sugar also, by storing it in warm chambers, so that the whole paste possesses a uniform temperature, lowering of temperature in the melangeur is avoided, and there is consequently no waste of the heating steam.

In some large factories the actual incorporation of cacao and sugar is preceded by a preliminary mixing of large quantities, which considerably relieves the strain on the melangeur, whilst it keeps the machine rooms as far as possible free from superfluous dust.

Fig. 28.

The mixing machine shown in fig. 28 can here be used with advantage. As will be seen on comparing the illustration, it is provided with a shifting trough. Such a

180

machine, when closed down, is capable of mixing from 100-500 kilos of chocolate. The mixing is effected by means of two suitably shaped blades, and the heating by a steam jacket. After the operation is completed, the mixed material is turned out into portable troughs, and after having been kept in a warm chamber for some length of time, transferred to the melangeur for further treatment.

It has been found advantageous to keep the chocolate mass so obtained in suitable receptacles for several days[114], at a temperature of not less than 20° C. and between that figure and 40° C. So the sugar is enabled to penetrate the entire mass, which now proceeds to the rolling processes carried out in the melangeur and rolling machines. Shortly before its discharge from the latter, it is mixed with spices, vanillin, eatherial oils and so forth.

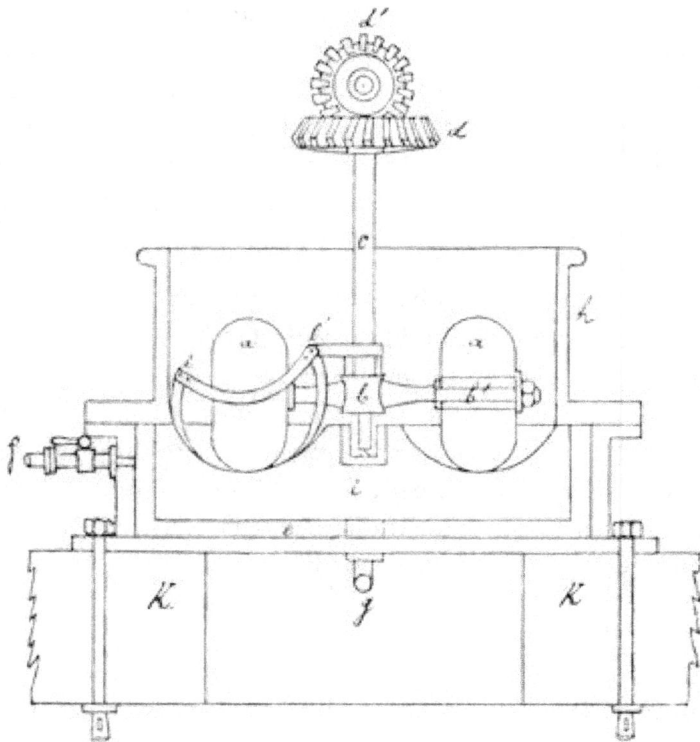

Fig. 29.

7. Treatment of the Mixture.

a) Trituration.

In describing the mixing machines, we do not intend to enter into details regarding the machines formerly in use, but merely to give a brief outline of the principles illustrated in their construction.

Trituration was formerly produced;

1. by rollers running backwards and forwards on a

grinder;

2. by several cones rotating in a circle on a disc-shaped bed;
3. by means of rotating stones running in a trough;
4. by means of several cylindrical rollers;
5. by means of grooved cone moving in a grooved casing.[115]

At the present time only the type mentioned under 1. and 4. are in general use. 5. is met with less frequently, and will be described at greater length in a subsequent paragraph.

The machines 1. and 3. are put into operation prior to the cylinder rolling mills, which finish off the incorporation of chocolate and sugar and the levigation process only begun in the first-named.

The machines constructed in the manner described under 3., to which we now turn, were introduced by G. Hermann of Paris, but are at present almost obsolete. Since they have some historical interest and are typical of the development of the melangeur, we annex a rough sketch showing their general construction in fig. 29.

The ellipsoid runners a made of granite work in the trough i which is also of granite and is fitted with the casing h. The runners rotate on their axles b so as to move in a circle. The two arms of the axis b have at the centre an elliptical ring with a quadrangular opening, into which fits the similar shaped part of the vertical shaft c fitted with the toothed wheels, d and d', which are set in motion by power transmitted to the shaft and its connections. The arm b has some play downwards. so that it can adjust itself vertically according to the greater or less quantity of material in the mill. The two steel blades, e and e', are shaped to fit the cavity of the trough; being connected with the shaft c they revolve with it and sweep down the cacao mass adhering to the sides of the mill. Between the foundation k and the trough *i* there is a space *l* into which steam can be introduced through *f*, the condensed water passing away by *g*.

All machines of this kind have now been displaced by the melangeur which is capable of turning out a much larger quantity of material with a relatively smaller expenditure of power. The operation of mixing chocolate is not a mere mixing, for the p r e s s u r e exerted by the r u n n e r s is also an indispensable factor. On that account the ordinary mixing machines have not proved serviceable, especially in the case of chocolates containing a small amount of fat, such as the cheaper kinds, while the addition of cacao butter to facilitate the working of the machine would considerably increase the cost of production. Melangeurs are generally constructed on the same principle as the edge runner grinding mills which are so much used; but they differ from them in so far as the bed-stone revolves, while the runners merely rotate on their axles without revolving.

Fig. 30.

The melangeur with travelling bed-stone, as constructed by Lehmann, is shown at fig. 30; it is fitted with an arrangement for lifting out the runners.

Fig. 31.

The bed-stone as well as the runners are made of granite. Each runner has an axis working in plummer blocks, so that it can be lifted out independently of the other one. By that construction the runners are prevented from taking an oblique position as was the case with the mills formerly made, since one runner would be forced downwards or tilted on its outer edge whenever the other one was raised up somewhat. The bed-stone of this machine revolves and it is easily heated by steam pipes from below. One important advantage of this machine is that being low it can be very easily charged and emptied. The contrivance for lifting out the runners prevents them thumping upon the bed-stone that might otherwise readily happen when starting the machine, and it also lessens the wear of the driving bands; moreover, large lumps of sugar or cacao are very readily crushed down and, so, the working is much facilitated. The emptying of the melangeur is readily and safely effected, while the bed stone is revolving, by holding a shovel so that the cacao is thrown up against the shovel. A melangeur of this construction is represented by fig. 31; it has three runners and underneath the bed-stone is fitted a steam engine which supplies driving power, the exhaust steam being used for heating the machine.

Although this emptying by hand is not attended with any serious drawbacks, yet it involves loss of time and is rather inconvenient, so that the demand for mechanical automatism in this operation was very considerable. It is now some years since Messrs. J. M. Lehmann patented an apparatus for the mechanical discharging of the chocolate mass from these machines, but their invention still holds good. A melangeur provided with such apparatus is shown on fig. 31. Here a vertically moving shovel is sunk behind the outlet, gradually damming the material, and causing it to rise above the edge of the tank and fall through the opening. A second but horizontally working arrangement, which in this case as in the last is controllable by means of a crank, conducts the remainder of the material to the same shovel. So the material is discharged within a few minutes. —These melangeurs are built for varying outputs. Fig. 32 illustrates one of the largest yet constructed. Its base has a diameter of 2 metres, and the machine itself has a capacity of 5 cwts. To avoid the mixing of dust with the sugar as far as possible, the whole melangeur is provided with a dust-proof protector.

b) Levigation.

An extreme fineness and homogeneity of the chocolate mass is obtained in the employment of cylindrical rolling machines, for the construction of which we are indebted to G. Hermann of Paris. Every kind of chocolate must be passed through the rolling machine at least once or twice even when finely powdered sugar is used, though in this case it is less a question of sub-division than of incorporation and intermixture. The best qualities are passed through the machine from six to eight times, or even more. The mass is finally fed into the machine in cold blocks and so ground off. Granite is the material chiefly employed in making the rollers, although it is not every variety which can be adapted to this purpose. Apart from the fact that granite, or indeed any other mineral stone, seldom occurs in compact masses and free from flaws, neither porphyry nor the stone generally described as granite is suitable for employment in the construction of mill rollers. A kind is generally preferred which intermediates between granite and porphyry as to hardness and possesses excellent grinding capacities, and which goes by the name of diorite. No other stone can compare with this diorite in respect to the above qualities, and the chief firms engaged in the construction of roller machines possess their own quarries. But we shall return to this later, for recently experiments with case-hardened casting rollers (Krupp steel) and hard porcelain have yielded very flattering results.

Fig. 32.

Fig. 33 a.

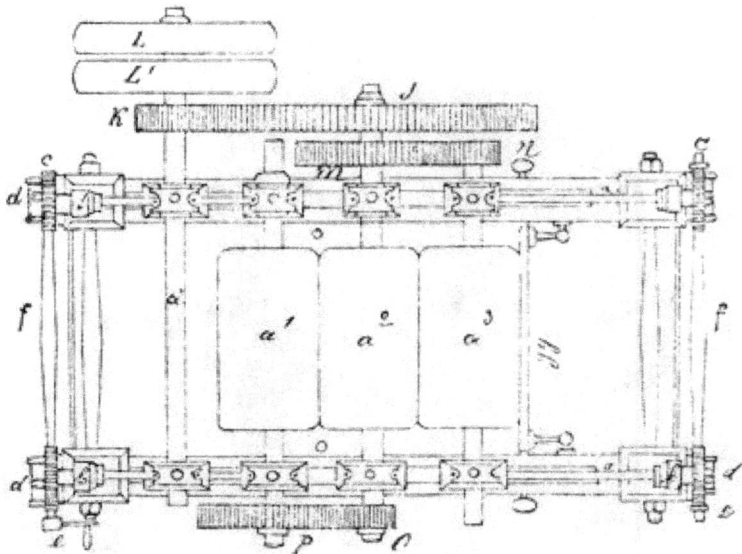

Fig. 33 b.

We shall now enter into more detail respecting the principle illustrated by these rolling machines. The plasticity of the chocolate mass necessitates a rotation of the cylinder surfaces in opposite directions with dissimilar velocities. Accordingly two or more rollers are caused to work against each other, and in compliance with this principle of sub-division with differential velocities, their axles are fitted with wheels, of which each has a different number of cogs.

So those rollers furnished with the greater number of teeth revolve more slowly, whilst in opposite instances there is a corresponding acceleration.

Fig. 34.

The construction of the machines now in use differs more or less from that of the type first invented by Hermann, plan and elevation of which appear in figs. 33 a and 33 b respectively.[116]

The granite rollers at a^1, a^2, a^3 are fitted with an octagonal iron axle that is somewhat thicker at the interior part and they are mounted upon a frame as shown in the drawing. The sockets of the central rollers a^2 are fixed and each one is held in position by three sets screws; those of the two other rollers can be shifted along grooves in the frame and when the cylinders a^1 and a^3 have been brought into proper position relatively to the cylinder a^2 they are held fast by the set screws p.

Fig. 35.

Fig. 36.

For the purpose of this adjustment, there is at each end of
the machine a horizontal wrought iron shaft f that can be
turned by the winch e, and these shafts are fitted with two
endless screws d working in the corresponding wheels c.
These occur on the spindles a, which screw in and out of

the bearing blocks of the rollers a^1 and a^2, but turn only in the fixed collars b without being shifted from their place. The result is that on turning the cranks e the corresponding cylinder a^1 or a^3 is moved nearer to, or further from, the central cylinder a^2, while the position of all of them always remains parallel. The shaft Q is set in motion by the driving wheel L fitted with the loose wheel L^1. It acts first upon the cog wheel K which works in the larger wheel J on the axle of the central roller a^2. That works in the cog wheel O and the wheel P fitted to the roller a^1 driving them as well as the wheel M and the pinion N of the roller a^3 The result is that the axle a^2 makes 1¾ revolutions and a^3 6-1/8 revolutions while a^1 in the same time makes only one revolution.

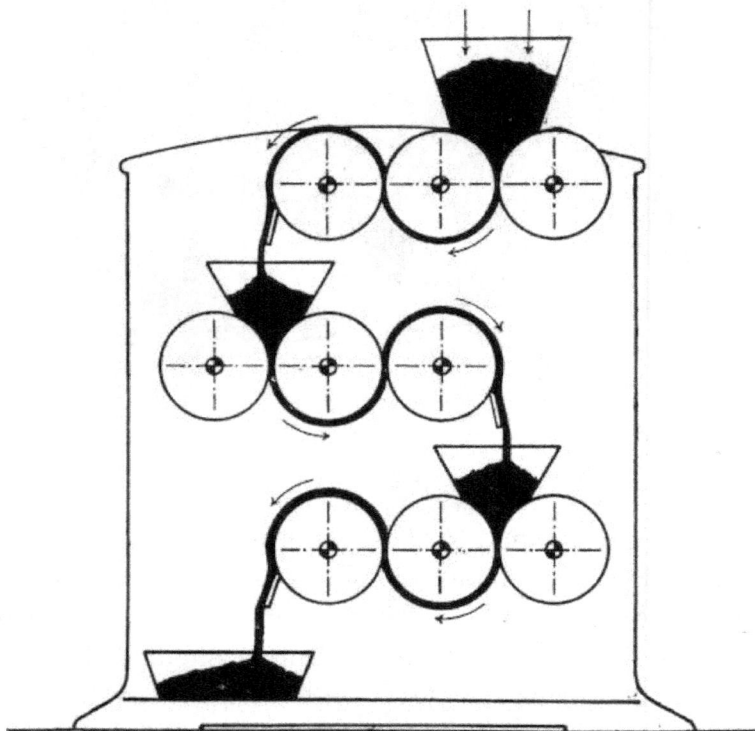

Fig. 37.

The cacao or chocolate is supplied to this machine by the hopper R which is placed between the rollers a^1 and a^2. The pasty mass adhering to the rollers is carried forward by the quicker moving roller a^2 and it is ground finer between the rollers a^2 and a^3, after which the material is removed from a^3 at the outer side by an adjustable blade gg and then falls down into a receptacle below.

J. M. LEHMANN
DRESDEN·PARIS·NEW-YORK.

J.M.LEHMANN
DRESDEN-L.

J.M.LEHMANN
DRESDEN-L.

Figs. 38 and 39.

On the design fig. 3434 we see a machine of more modern construction ready mounted. The receptacle parts of the same are arranged and connected in full agreement with the above mentioned except that the motion is effected by the driving power fitted to the machine on the ground on the left side.

Fig. 40.

The principle of this roller machine has long been applied in the building of other types, and we find that these, variously altered, renovated and improved, are to-day an indispensable equipment in every chocolate factory. In the following pages we give a description of some of the best-known constructions of refiner.

The so called battery rolling mills constitute a remarkable innovation. It is apparent that the more rollers a cylinder machine contains, and the greater their length and diameter, all the more efficacious will the working of the machine be. Batteries have accordingly been constructed, whereby two, three or more roller systems are combined, one to every three rollers, and rising one above the other, so that they slant upwards much as shown in Fig. 40.

As the battery rolling mills possessed the disadvantages that they took up too much room and could not be well fed and regulated, they are generally replaced by rolling machines of from 6 to 9 rollers, first constructed by J. M. Lehmann. These rolling machines of 6 to 9 rollers which we see before us in Figs. 3535 and 3636 are really systems of 3 rollers fitted one over the other. They therefore take up the room of a 3-roller machine and are quite as easy to work.

Fig. 41 a.

As will be seen from the design of a nine-roller apparatus, fig. 37, the chocolate mass descends from one roller system to the other, and is fine rolled in a third of the time otherwise required, and at one operation, with corresponding saving of labour. The nine roller apparatus are provided with landing stage and steps, and fed either by means of elevators, or from above.

Fig. 38 shows a recent construction, three roller apparatus (case castings, cf. below) standing vertically, which accordingly takes up little room. The hopper is low-lying, whilst the discharging is effected from the upper roller, and accordingly admits of the occasional use of a somewhat larger size of transport trolley. This type also occurs with 6-9-12 rollers, as apparent from fig. 39.

Fig. 41 b.

Whilst these systems were exclusively supplied with rollers made of granite or hard porcelain up to a few years ago, it has been found that good results are obtained by the use of cast rollers, and they have been for some time employed on machines of three, four and five rollers. (Figs. 38, 40, 41 a-c.) In consequence of the non-porous surface of these steel rollers, it is possible to grind to a finest powder, merely in one operation, without passing the chocolate through the machine several times; and the so-called "Burning" of masses which have not been properly mixed cannot arise in this case, though it is true that the apparatus must be provided with water cooling arrangements to avoid a too excessive heat. They are specially adapted to the preparation of the more ordinary qualities, and are even occasionally employed for finer chocolates, for obviously these must be again submitted to a rolling process, when granite or porcelain rollers are preferred.

Fig. 41 c.

For this reason the 6, 9 or even 12 roller mills have been more discarded since the last grinding process has been performed by granite rollers (cold process).

In order to avoid the disadvantages of the pulley drive, it is in certain cases advisable to drive each machine direct from an electric motor. Fig. 42 illustrates a refining machine driven in this manner.

Fig. 42.

c) Proportions for mixing cacao mass, sugar and spices.

The relative proportions of cacao, sugar, and spices, as well as of starch as in the manufacture of the cheaper sorts, vary considerably. Generally speaking 50 or 60 parts of sugar are added for 50 or 40 parts of cacao mass; the following are a few formulae applicable for the production of those kinds of pure chocolate that are most used.

A. Hygienic chocolate.

Cacao mass
Powdered sugar } equal parts of each.

B. Spiced chocolate.

a)

Cacao mass	4 kg
Sugar	6 kg
Cinnamon	72 g
Cloves	38 g
Cardamoms	16 g

b)

Cacao mass	4 kg
Sugar	6 kg
Cinnamon	130 g
Coriander	8 g
Cloves	88 g
Oil of lemons	2 g
Cardamoms	16 g

c)

Cacao mass	5 kg
Sugar	5 kg
Cloves	80 g
Cinnamon	220 g
Mace	8 g

d)

Cacao mass	5 kg
Sugar	5 kg
Cinnamon	100 g
Vanilla	100 g
or Vanillin	2·5 g
Mace	2 g
Cardamoms	4·2 g

e) Spanish spiced chocolate.

Cacao mass	5 kg	Cardamoms	82 g
Sugar	5 kg	Mace	44 g
Cinnamon	116 g	Vanilla	40 g
Cloves	50 g	or Vanillin	1 g
		Oil of lemons	1 g

C. Vanilla chocolates.

a)

Cacao mass   5 kg

Sugar	5 kg
Cinnamon	160 g
Vanilla	50 g
(or Vanillin	1·2 g)

b)		c)	
Cacao mass	4½ kg	Cacao mass	4 kg
Sugar	5½ kg	Sugar	6 kg
Cinnamon	150 g	Cinnamon	120 g
Vanillin	1·5 g	Cloves	20 g
		Vanillin	1·6 g

The powdered spices as given above may be replaced by corresponding essential oils, but see page 237 for remarks on this point.

If the chocolates made from beans rich in oil contain too much fat to mould properly, a small percentage of their constituent cacao mass can be replaced by cocoa powder made from the same kind of bean, but defatted, in the case of the finer qualities; and when inferior varieties are under consideration, the same result may be attained by a sufficient increase in the proportion of their other constituent, sugar, as e. g. 55-60 parts of to 45 or 40 parts of cacao mass, so disturbing the usual equality of the two ingredients mixed together. Very cheap chocolates in particular are prepared from a smaller percentage of cacao mass and show a corresponding increase in their sugar content. But if the sugar exceeds 65 percent, it is no longer possible to mould these chocolates, and the addition of fresh cacao butter becomes a necessary preliminary to this operation, cf. also the first part in section IV. Such varieties would have a composition somewhat like the following:

| Cacao mass | 25 parts |

Sugar	67 "
Cacao butter	7 "
Spices and vanillin as above	1 "

In the experimental preparation of samples of chocolate mass it is not advisable to employ large quantities of ingredients, when a waste of material is bound to ensue, but to begin with mixing small quantities of one or two kilos. The small Universal Kneading and Mixing Machines, Patent Werner & Pfleiderer, Type 1, Class BS, can here be used with advantage. They are specially intended for small outputs and experimental work; but we shall return to their description later, after stating that they are furnished with heating apparatus, stuffing boxes and air-tight lid, and can easily be taken to pieces, greatly facilitating the removing of the mass.

III. Further Treatment of the Raw Chocolate.

8. Manufacture of "Chocolats Fondants

Fig. 43.

Recently the creams sometimes described as in the heading have enjoyed a vast popularity, and are sold as eatable chocolates in ever-increasing quantities. As far as can be ascertained, they were first manufactured in Switzerland, melt readily, and have a correspondingly large amount of fat, resulting from the addition of cacao butter, which distinguishes them from ordinary chocolates. When readily melting chocolates were first introduced, it was a prevailing opinion that the required property could only be obtained by increasing the amount of fatty content. Now the excessive evidence of fat in chocolates is very objectionable, both as regards taste and digestibility. To avoid this, therefore, the chocolates are treated mechanically, to attain the required character of readily melting. The machines used for that purpose are termed "Conches", because the trough, in which the chocolate is rubbed into a long cylinder, has somewhat the shape of a long shell. For the working up of chocolates in conches, the necessary conditions are;

1. that the chocolate should have been ground perfectly fine,

2. it must contain such an amount of fat as to become glucose on warming, not indeed so thin as that used as coating material, but nevertheless softer than the ordinary cake-chocolate of good quality.

Fig. 43 a.

Fig. 44.

The machine can be heated by means of steam, hot water pipes, gas or charcoal stoking, according as they are available in the place of installation, and the temperature should rise above 70-80° C. for fondants, and 50° C. for milk chocolates. In factories with water power or electricity, continuous fondant machines can be worked day and night, but when only worked during the day, must be kept warm overnight. Constant tending of the machine is unnecessary, as it works automatically. After a treatment of from 40-48 hours, the chocolate attains the requisite character (i.e. it melts readily), and a rounding off of taste, which are the properties of all good brands Milk chocolates can also be advantageously prepared in the conche, as also covering or coating cacaos of all kinds, which harden considerably in consequence of this treatment.

Figs. 43 and 43 a show quadruple conches of the modern type with hot water, wherein four troughs are arranged in pairs, and one opposite the other. Conches with only 1 and 2 troughs are also constructed, and in various sizes, the troughs sometimes having a capacity of 125 and 200 kilos, so that the quadruple conche is capable of holding five or eight hundred kilograms in all. The curved bottom of the troughs, as well as the rollers fitted in them, are made of granite, and the front wall strongly bent in at the corner, so that the mass is forced over the border of the front wall, where there are openings for its discharge as well. To prevent radiation as far as possible, it is best that the troughs be walled in, the troughs are either walled. Fig. 44 shows the room of a modern chocolate factory, with 15 conches.

"Chocolats fondants" are from a gastronomic point of view, the finest chocolate product on the market, and it is not remarkable that this branch of the chocolate manufacture has witnessed a considerable extension, and is likely to extend still more.

9. Heating chambers and closets.

The manufacture of chocolate has been very considerably facilitated by the introduction of heating chambers and closets, which have now become an indispensable feature of every factory in the industry. In these chambers the chocolate which has still to be rolled, as well as that already submitted to this operation, is stored and kept at a temperature of 60° C. until it can be further treated (moulded). This manner of heating involves an appreciable cheapening of the production, for masses which are dry and apparently require an addition of fat recover in such a manner during a twenty four hours' storing in the heating chambers that such addition becomes unnecessary. But especially when chill casting rollers are employed, which the mass leaves in a very dry state, the use of these heating chambers is indispensable. They should be available in every factory to such an extent as to find room for the total output of one day, though even twice or three times this amount might very well be provided for. Closets heated by steam are best adapted for small factories, such as are illustrated in Fig. 45. They possess double doors, are walled in, and are capable of holding from 300-400 kilos of chocolate mass for each metre of length. Larger factories should furnish themselves with chambers, which are more open to access and on the walls of which iron shelves can be introduced, heated by steam pipes arranged underneath. A typical chamber, measuring 2·8 metres in breadth (including passage) and 5 metres in length would hold about 2,500 kg of chocolate.

Fig. 45.

10. Removal of Air and Division.

Fig. 46.

After emerging from the final rolling process, the chocolate is stored up in heating chambers until it is ready to succeed to the moulding, prior to which, however, it must be freed from air and cut up into small portions. Until recently, it usually came next in a melangeur provided with a dish-shaped bed-stone made of granite, as illustrated in fig. 46, where it was kneaded and reduced to a uniform plasticity and heated to the temperature required for moulding. The melangeurs devoted to this purpose are now superseded by special tempering machines.

A machine of this recent construction, used for working solid and semi-liquid material, is shown in fig. 47. The tank intended as a receptacle for the chocolate mass is in this case made of iron and, to facilitate cleaning, smooth in the interior. It runs in a water-bath, the supply in which can be controlled by steam or cold water. The granite runner is provided with a lifting device, admitting of the working up of material containing foreign ingredients like nuts, whole and fine-split.

Fig. 47.

The mass is taken out of the machine in lumps, and in order that it may be reduced to a temperature suitable for the removal of air (about 26-32 ° C. on the outside) it is laid to cool on wooden, marble or iron tables. When this temperature is arrived at, large lumps of chocolate are introduced either into the air-extracting or the dividing machines.

Fig. 48.

After the importance of the tempering processes had at length been recognised, inasmuch as the maintenance of the temperature prescribed is of immense influence on the chocolate subsequently produced, and it had on the other hand been ascertained that such machines as described above could not be absolutely relied upon, for the shaking tables involve an occasional excess of tempering, the idea of a machine which should completely and automatically perform this task was finally conceived. This new machine, given in fig. 48, and already differing from all other tempering machines in external appearance, ushers in an entirely fresh process respecting the *modus operandi* prevailing in the present manufacture of chocolate, which does not fail to satisfy the highest expectations. It may be said to work continuously, for no matter what the temperature of chocolate passing into it may be, the material leaves the machine at the temperature desired within a lapse of one minute. According as more or less chocolate has to be turned over in the moulding department, the machine can be stopped or set in motion without detriment to its efficacy. Besides this, it cleans almost automatically, so that a quick change of quality is always possible. The special virtue of this machine is that it turns out the material with such a degree of homogeneity as has never before been known, making moulding at much higher temperatures a possibility. There is yet another side issue, namely a doubling of the life of the moulds, and finally, owing to the fact that the often considerable amount of waste material is done away with in this process, the moulding shop is spared to some extent. The series of rollers through which the chocolate passes is maintained at a proper temperature by means of automatic water apparatus. The daily output of the machine figures at 3000-4000 kilograms. The material is passed on out of this machine to the dividing and moulding processes.

The necessary extraction of air follows immediately on the tempering process, for the blades of the scraper then release the chocolate mass from the rollers in thin layers, between which air penetrates. The removal of air is effected by machines, an old type of which is shown in fig. 49 (in front elevation).

It can be warmed by means of a charcoal fire placed in the space **i**, or by any other suitable means. The chocolate mass is fed into the cylindrical hopper **a**, at the base of which occurs an archimedian screw **b**, which is propelled by the shaft and cog-wheel system **c d e** in the direction indicated by the arrow. Thus the chocolate mass is forced into the box **f**, leaving which in cylindrical form, it succeeds to the travelling band **h**. It is now almost entirely freed from air. As the material is pushed forward on the band, it is cut off either by a knife **g** fixed to the box **f**, or divided as far as possible into equal parts by a double knife with adjustable blades corresponding to the weight required for a chocolate square. This manipulation presupposes a fair amount of skill on the part of the machinist, but this once attained, the division ensues as precisely and simply as can be desired.

Air-extracting machines of recent construction, although still partially built on the above principle, are at the same time generally developed as automatic dividing machines.

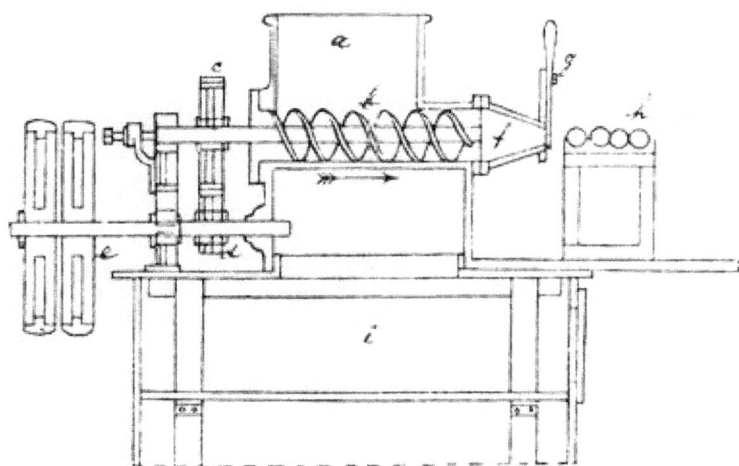

Fig. 49.

Fig. 50.

Fig. 50 shows such a machine for solid and semi-liquid chocolates. By means of this, the material is next conducted along a vertical screw path in even mass to the horizontal screw, and so a second filling with the hand is rendered unnecessary. After it has been freed from air in this, it enters a revolverlike cutter, which discharges the divided portions on a travelling belt. On the latter it is conducted to a table standing near, where it is laid into moulds. The machine is of very strong make, and puts out from 15-250 gr, divided into approximately 10-25000 squares, within a space of ten hours.

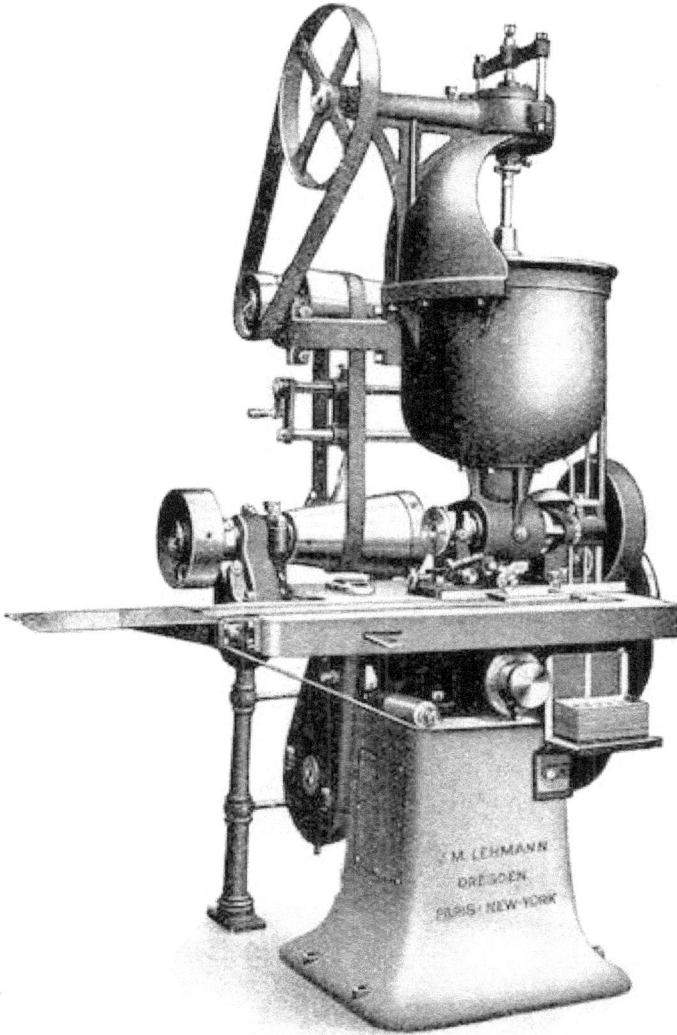

Fig. 51 a.

Figs. 51 a and b give finally two of the best known types
which have a very extensive application, protected by patent
imperial (Germany), and built by J. M. Lehmann, Herm.

217

Baumeister, J. S. Petzholdt in Döhlen, G, near Dresden and others. With this patent dividing machine of J. M. Lehmann, solid and semi-liquid chocolate material, as also nut and almond chocolates are divided exactly, in any weight from 18 to 250 grammes, and then conducted in strips of equal size to the mould previously mentioned. As far as cleanliness, purity, and easy management are concerned, it fulfills all the demands which can be expected of the most modern machine.

Fig. 51 b.

IV. Moulding of the Chocolate.

11. Transference to the Moulds.

Fig. 52.

The pieces of chocolate, or emerging from the dividing processes, are placed separately in iron moulds, that is to say, as far as this has not already been done in the dividing machine. It is important that these should have the same temperature as the chocolate mass, in order to prevent the formation of spots on the surface of the cakes, and to obtain a good and non-greyish fracture. The temperature for moulding smaller objects can be fixed at between 27° and 32°C. and for the larger may be considerably lower. In summer also, moulding may be proceeded with at a lower temperature than in winter. According to a note in the Gordian (1895, No. 4) the moulding may be carried out in summer, when the atmospheric temperature is;

from 25-31° C, at 26-27·5° C
" 18·5-25° C, at 28·5-30° C
" 12·5-18·5° C, at 31-32·5° C

219

In cold weather, the cakes may be moulded at a temperature of from 32·5-35° C., according to their thickness. When not manufactured in the automatic machine shown in fig. 48, the mass should be otherwise controlled as regards temperature, which should be registered by a thermometer introduced therein. The moulds are for the most part filled with plastic and liquid chocolates, and their depths determined and modified by the weight of material which they are destined to receive.

Fig. 52 shows a machine which conducts the semi-fluid mass to the moulds in the following manner. The moulds are automatically introduced under the apparatus, and filled from the small stirrer above. They then succeed to the shaking table and are finally transported to the cooling room. On this machine moulds of from 75-350 mm long and 75-225 mm broad can safely be employed.

There are two different forms in which chocolate is sold, namely, that intended for domestic purposes, and that which is to be consumed as an article of luxury. The kinds known as cake, rock and roll chocolate belong to the first class, the several pieces weighing 50, 100, 200, 250, 500 up to 5000 grammes. Tin-plate is the only material of which moulds are made; and these generally have a capacity rather greater than is necessary for holding the particular quantities to be moulded. The chocolate is therefore, as described above, divided into given weights, and generally deposited direct in the moulds by the dividing machine. The divided portions of chocolate are pressed down in the moulds by hand, equally distributed in the latter, and then transferred in the moulds to the shaking table or combination of shaking tables to be described later. On the shaking table the soft chocolate soon penetrates completely into all the corners and impressions of the stamped tin moulds. The removal of the cooled cakes from the moulds is easily effected by pressing.

These moulds are generally provided with from four to ten ridges or indentations, so that the chocolate can be conveniently divided, and as required for use. Others again have a similar number of compartments.

The compartments may be impressed with any kind of inscription, so that such information as the name of a firm can always be reproduced on the cakes.

Broken chocolate is generally of inferior quality, brought on the market without any protective covering.

In those kinds of chocolate which are known as articles of luxury a distinction is to be made between;

1. Those moulds which are in one piece and completely filled with chocolate, so that the superfluous mass can be removed by a knife. In such cases the weight of the cakes is exactly regulated by the capacity of the moulds.

2. Those intended for moulding various figures of fruit etc. in which two or three parts make a closed space which is of the form desired.

Among the moulds of the first type must be numbered those used in the preparation of small tablets and sticks, and the sweetmeats known as Napolitains and Croquettes.

The second class comprises moulds for making chocolate cigars and chocolate eggs, and also the double moulds.

The moulds for the smaller tablets, cream sticks, napolitains and croquettes are also made exclusively from tin-plate, and the separate parts are enclosed in a stout iron frame, the top of which is ground down smooth, so that any superfluous portion of the filling can easily be scraped away. In that way from six to thirty pieces can be cast in one mould at the same time: the cooled chocolate can be released from the moulds by gently tapping one corner against a table. In napolitain moulds protecting hooks are attached, to avoid their sustaining any injury in this operation.

Examples of the more frequent moulds.

1. Chocolate Cigars.

These are made either by introducing the chocolate mass between the two halves of a double mould, of which each corresponds to a half of the cigar shape to be moulded and which each fit exactly one on the other; or else by pouring it into hollow moulds stamped out of one complete piece. Moulding presses[117] are utilised in the manufacture of material en masse In these the cigars are filled into iron moulds, afterwards held together by means of iron combs, and so introduced in to the press. For each size and shape special moulds and plates are essential. Neither barium sulphate nor zinc white may be employed to produce an imitation of the ash on ordinary cigars, as both are objected to by health inspectors; nor are they necessary, for in phosphate of lime (tricalcium carbonate) we possess a perfectly harmless and at the same time efficient substitute, when it is mixed up with starch syrup.

Other figures, such as fish etc., may also be produced in chocolate, by means of the moulding press, when it is furnished with stamped moulds, corresponding to the forms required.

2. Chocolate eggs.

These are generally made hollow, unless they are very

small, by pressing chocolate in two halves of an egg-shaped mould and then uniting the two parts. Another method patented by Th. Berger of Hamburg[118] seems less practical. A mould is made of soft sheet caoutchouc blown out; this is dipped into liquid chocolate and, after the adhering coating has hardened, the air is let out of the mould. The use of caoutchouc moulds would render this method too costly, since the alternation of temperature soon makes the caoutchouc unserviceable.

3. Various figures, fruits, animals, and other small objects.

Double moulds are used for making these objects in chocolate, consisting sometimes of three or four parts; they are made either of sheet iron, tinned, or, for more complicated forms, the moulds are cast in tin, but these latter are not so durable as those of tinned sheet iron with strong iron frames.

The several parts of the moulds, after having soft chocolate mass pressed into them, are put together and excess of material is removed by requisite pressure by the use of a press of the kind made by A. Reiche in Dresden, which will admit of a large number of moulds being placed in it at a time. By the use of such a press the moulds are protected from injury, and the objects moulded have a better appearance, as a result of the uniform as well as strong pressure exerted.

After cooling, the moulded objects are readily detached from the moulds and they only require to be scraped clean, or further ornamented as may be desired. That is done in various ways, for example by painting with coloured cacao butter.

4. Crumb Chocolate.

This term is applied to the small pieces of chocolate of truncated conical shape, with from 4 to 5 smooth surfaces. They are made by a machine specially constructed for the purpose by A. Reiche (No. 1550); it consists of a four-cornered box with a removable bottom. Inside the box there is a false bottom, from 1 to 2 cm above the other bottom, which is fitted with a removable sheet iron plate, in which pentagonal holes are stamped. A knife can be introduced at one corner of the bottom of the box. After sufficient chocolate has been made to penetrate through the pentagonal holes by agitating the box on the table, the knife is rapidly drawn across the bottom and the box raised up. The sheet iron plate is then taken out, and by gently tapping one corner the small pieces of chocolate are shaken out.

5. Small tablets, sticks, fruits or figures filled with cream.

These are prepared by pouring the cream contents in either wooden or iron moulds, previously dusted with a little flour, and then moulding round them chocolate in whatever form is required, always taking care that this is kept as soft and plastic as possible, a suitable addition of cacao butter proving invaluable for the purpose.

In former times chocolate moulds were manufactured exclusively in France, where the firm Létang of Paris enjoyed what was to all intents and purposes a world monopoly. But since the year 1870 the oft-mentioned firm of Reiche in Dresden-Plauen has taken up the manufacture, and has succeeded in conquering the market in a remarkably short time. The moulds of this firm satisfy each and every possible requirement, although it would be no disadvantage if the old type of pattern mould were cleared away at one and the same time with the old routine, to make room for a little artistic skill and embellishment.

Recently Reiche has brought out a special machine intended as an easy and practical cleanser of his many moulds, which include bonbon cutters and cutting rollers, numbering stamps, chocolate slicers, roller machine boxes etc. He has lately brought on to the market a special machine for quickly and efficiently cleaning the moulds, which is illustrated in fig. 53.

In one end, a circular brush is introduced, and against this the moulds to be cleaned are firmly pressed. In consequence of the large number of revolutions which this brush passes through, the moulds are cleansed of still adhering masses of chocolate in a half or third of the time occupied when hand labour is employed. At the other end of the shaft occurs a duster, sprinkled with Vienna white (a lime), which polishes off the moulds previously and thoroughly cleaned by the circular brush. The great advantage of this machine is that the daily expenditure on polishing is considerably reduced One girl can do the work of two hand workers, when this machine is employed. In addition, it makes possible a continual touching up of the material used in the making of the moulds, a ventilating apparatus removing all traces of dust.

Fig. 53.

12. The Shaking Table.

The pasty chocolate mass fills itself into the chocolate moulds spontaneously, in consequence of its soft consistency. Yet to share it evenly throughout the mould, so that it adapts itself to every bend and hollow there occurring, and further to bring to the surface any possible bubbles of air evident in the mass, the chocolate is whilst still in the moulds subjected to brisk shaking.

This is effected by placing the chocolate on trays and transferring these to the shaking table, of which types and construction are at the present time manifold and various, the best and oldest being given in front elevation below (Fig. 54).

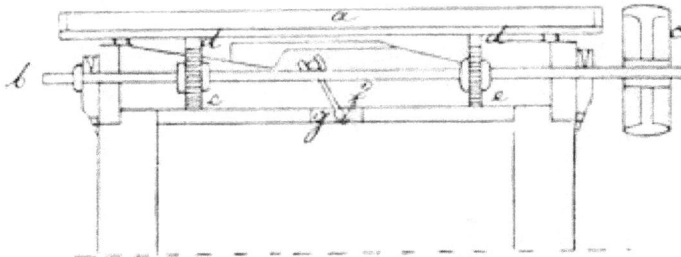

Fig. 54.

The movable slab **a**, fitted with an upright rim at its edges, has underneath two projecting pieces **d**, working against deeply toothed wheels **e**, which fastened on the shaft **b**, are driven round by the pulleys **c**. The teeth of the wheels catch on the projecting pieces at every revolution of the shaft and push them rather gently on one side, and when the tooth-points slide from under the slab, it drops down as much as it has been previously raised. Each tooth of the wheel coming into contact with the projections, the same motion is repeated several times, causing the slab to oscillate up and down.

This oscillation of the slab is controlled by means of a hand lever **f**, occurring on the shaft **g**, and fixed crosswise thereon, so that we can only show it in cross section on the diagram. The lever **f** attaches itself to the under part of the slab, raises it, and so throws the wheels out of contact with the projecting pieces, but without stopping the rotation of the shaft **b**.

Shaking tables have also witnessed considerable improvements with the lapse of time, and we shall now proceed to treat these in more detail, especially as several recent constructions offer and illustrate many interesting mechanical points.

Fig. 55.

Fig. 56.

An old type of machine, that is nevertheless still much employed, is illustrated in fig. 55. Here the slab is caused to osculate by shaking wheels introduced underneath, each possessing six, eight, or more teeth. The slab is raised and lowered by contact with wedge-shaped parts, the effect produced being greater or less according as the moulds are large or small, heavy or light, and in proportion to the consistency of the chocolate mass which they contain, e. g. whether it is solid or semi-liquid.

Quite an improved construction is shown in fig. 56. Here the table is attached to a vertical axle, which is moved up and down by means of a toothed wheel fixed on its bottom end. There is also a cylinder arrangement under the whole machine to assist in controlling the vertical motion of the shaft, and as it is provided with automatic lubrication, there is no danger of any wearing out of the apparatus and consequent irregularity of functioning.

The shaking and jerking of the slabs is in itself attended by a considerable amount of noise, and when to this is added that caused by the tables, it will be seen that a chocolate factory may become to its neighbours a very serious source of objection. For years attempts have therefore been made to construct shaking tables, so that they would not cause any greater noise than is absolutely inevitable. Pneumatic contrivances and caoutchouc have met with right royal success in this connection.

Fig. 57.

The most recent and probably the most perfectly constructed shaking table is given in figs. 57 and 57 a. It embodies all the latest improvements and is self lubricating, a fact of the highest importance as releasing the strain on the attendance, which would need to be very perfect to ensure absence of noise in the case of a machine making 800 strokes a minute. When it is considered that the moulding room is generally managed by girls who neither possess knowledge of, nor interest in, the machines, the advantages of such automatic lubrication become even more apparent.

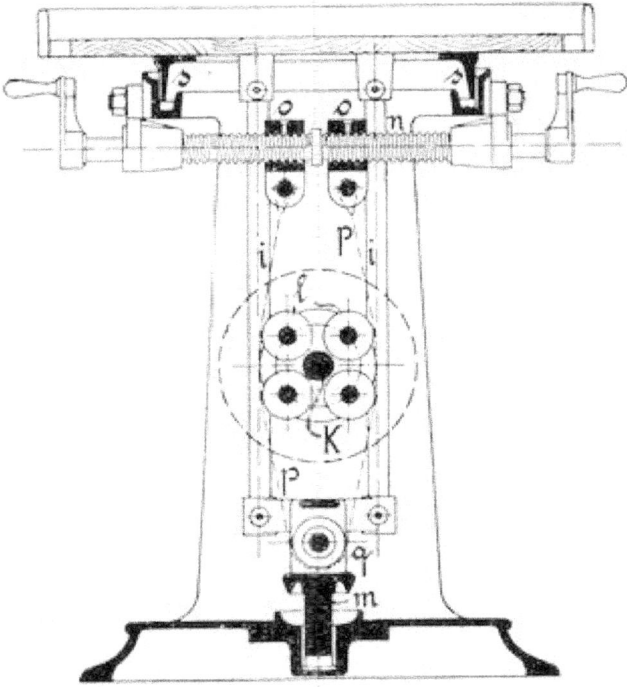

Fig. 57 a.

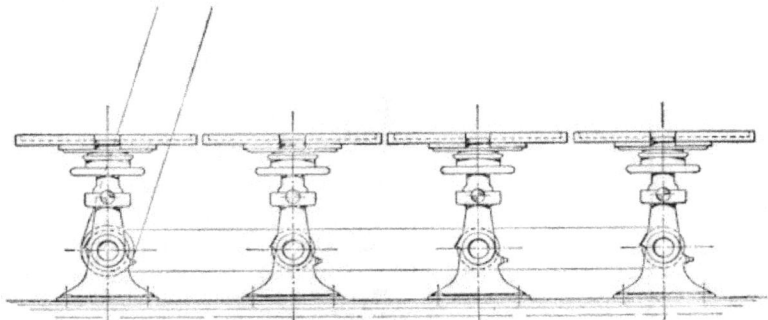

Figs. 58 and 58 a.

Apart from the automatic lubrication, in itself a sufficient
guarantee for the efficiency of the machine, screws and nuts
are entirely avoided on this machine. The motive mechanism
is also interesting. By a special arrangement, the number of
revolutions in relation to the number of the elevations of the
slab is reduced to one fourth, viz., from 760 to 190. Since the
elevation of the slab can be regulated to zero, a loose pulley
for shifting the driving belt is unnecessary; in addition, the
driving shaft makes only a small number of revolutions,
and works in oil. The round shaped upright serves to carry
the vertically moving frame **i**, which supports the slab
moving in an oiled groove at **s**, and which is supported
underneath by the pivot **m**. Both at **m** and **s** there is
automatic lubrication. The bearings of the spindle **n**,
attached to the upright, work into left and right screw
threads at **oo**, to which points the ends of a broad leather

belt **p** are attached, passing over the roller **g**, by which the frame **i** is suspended. The driving pulley **k**, running in oil, carries in its centre the four rollers **l**, which turn round and round the pulley **k**, so as to come into contact with the belt **p** and press it outwards on both sides. At the same time it shortens the belt in the vertical axle, so raising up the table slab **i**. This is repeated four times by one revolution of the driving pulley, so that working with 190 revolutions a minute, the slab is raised 760 times. According as the screws **oo** are moved to or from the centre, the vertical movement of the slab can be increased or decreased to a point when the slab is completely out of action, i. e. when the rollers l no longer touch the belt **p**. Under favourable local conditions, a number of such tables can be driven by one shaft, so that only one pulley and a single driving belt would be needed, though each table would work quite independently of the others. Such an arrangement is shown in figures 58 and 58 a.

Fig. 58 b.

This shaking table, though only recently introduced, has quickly made itself popular, and is especially suitable for the preparation of readily liquefiable chocolate. The gentle vibratory motion produced by this shaking table and its exact adjustability admit of the thinnest cakes being made in a perfectly uniform thickness, without any objectionable projections round their edges. Besides the shaking tables of

233

this construction there are others made in such a way that whether the moulds are light or heavy, small or large, the slab is always raised to the same height, the working of the slab being adjusted by altering the number of revolutions. The manipulation of these tables is much more difficult than that of tables constructed as above described, and that is probably the reason why these have for decades been scarce on the market.

The moulded chocolate spread out on trays is transferred as rapidly as possible to the cooling chamber, with which we shall conclude section IV.

Instead of several shaking tables alternately receiving the moulds, which involve frequent changes, so-called shaking systems (fig. 58b) have been generally adopted of late. They consist of a number of shaking tables, having their frames attached to each other, possessing a common motor control, and having their slabs arranged one after the other in such a way that the filled moulds slowly proceeding from the dividing machine can be automatically conducted over them. The shaken moulds are then passed on to further processes, or they enter the cooling chambers at once. The advantage of the shaking table system lies in a reduction of the number of hands, who only need to be in attendance at each end of the system, and further in the regularity, both as regards time and strength, which prevails in connection with the shaking of each mould.

13. Cooling the chocolate.

Experience has shown that the more rapidly the moulded chocolate is cooled the finer is its texture and the more uniform the appearance of the fractured surface. That is due to the formation of smaller crystals of the fat when the

234

cooling is rapid, while in slow cooling larger crystals are formed and the fracture consequently becomes dull and greyish.

Formerly it was possible to distinguish chocolate made in summer from that made in winter by the more uniform appearance of the fracture, that was, in the latter case, the result of more rapid cooling.

At present, however, manufacturers are no longer dependent upon favorable atmospheric conditions in that respect, for by suitable arrangements it is now possible to produce the reduced temperature requisite by artificial means.

The most suitable cooling chamber is an underground space which should, however, be so situated as to be in convenient communication with the moulding room. The cheapest and simplest place for a cooling chamber is a cellar, if it be properly constructed and dry, as well as large enough to contain the quantity of chocolate made in one day's working. The best temperature to be kept up ranges from 8° to 10° C. Within those limits there is no danger of the chocolate being coated with moisture, or that it will acquire a coarse grained texture by lying too long. The following rules will serve for guidance in regard to this point:

Generally, chocolate presents the finest fracture when it has been fully levigated and when it contains a considerable amount of fat, provided that the fat present is only cacao butter. Those kinds which are not so well levigated, or have had some addition of foreign fat of higher melting point, show an inferior fracture. It is possible to obtain an equally vitreous fracture in a less cold cellar (16° C. and upwards) when the chocolate is moulded at a temperature corresponding to that of the cellar; to effect that, the chocolate should be moulded at a proportionally lower temperature the warmer the cellar is. The difference can be seen by the appearance of pale red spots on the surface. When it is desired to dispense with artificial cooling, the cellar should be as much as possible below the surface of the ground; it should also be of sufficient height, not less than 3 m. If the situation and height of the cellar be properly adjusted, the requisite area for disposing of a daily production amounting to 5000 kilos would be 400 sq. m. The cellar must be well ventilated and furnished with double windows, so placed as to open towards the north and east. Discharges of warm waste water, as well as steam pipes or furnaces should be kept as far distant from the cellar as possible. The internal arrangement of the cellar should be of such a nature that the whole of the chocolate to be cooled can be deposited upon the floor, since that is the place where cooling takes place most rapidly. With that object in view it is desirable to construct brickwork pillars about 25 cm high, covered with white tiles. Passages are arranged between these pillars. The cellar should be entered by as few persons as possible and, therefore, the cooled cakes of chocolate should be taken at once, in the moulds, to an adjoining room to be turned out and passed on to the packing room and store.

Most of the existing factories, that have been established for any time (large and small) have had to adopt artificial means of cooling, because in most instances the quantity of chocolate to be cooled daily has, in course of time, increased tenfold. The machine rooms have been enlarged, the number of machines has also been increased, while the cooling cellar has remained in its formerly modest proportions. But those circumstances are not the only reasons for having recourse to artificial refrigeration, which is often necessary in consequence of the inconvenient situation of the cellar and the high underground water level.

In the application of artificial refrigeration in a chocolate factory it is not advisable to hasten the cooling of large quantities by producing too low a temperature in small chambers. The cakes of chocolate mass by that means come out of the moulds as hard as glass, but it is questionable whether the consumer using the chocolate many months afterwards, will make the same observation. Great care would have to be taken with such rapidly cooled chocolate, to pass it gradually through chambers of a medium temperature and thus prepare it for exposure in the packing rooms and warehouses. Even when employing artificial means for cooling, the reduction of the cellar temperature and cooling upon pillars is to be preferred to the more direct cooling upon a system of pipes, which after all is nothing else than a cooling upon ice, as may be in some instances the only alternative. Consequently, a well constructed cellar for cooling, furnished with a system of cooling pipes on the roof is perhaps the most advantageous arrangement, especially for large factories.

In carrying out artificial refrigeration various kinds of machines are used for reducing temperature, in which the desired effect is produced either directly by the condensation and evaporation of suitable materials, such as liquid carbonic acid, ammonia, sulphurous acid, or indirectly by making saline solutions (calcium chloride), cooled below the freezing point, circulate through a system of pipes fitted on the roof or walls of the space to be cooled. As the cold liquid is pumped through the pipes, it takes up heat from the air in contact with them, correspondingly reducing the temperature of the cooling chamber. The cooling installations of the firm of C. G. Haubold, junior, Chemnitz, are among the best and have long been extensively used in the chocolate industry. Their cooling apparatus is a compressing machine, in which coolness is obtained by the evaporation and recondensation of such liquid gases as carbonic acid or ammonia. Like all compressing machines, it is comprised of three main parts.

I. The evaporator or refrigerator, consisting of a wrought iron system of pipes. The latter are placed in the spaces of the plant to be cooled, with a so-called direct evaporation arrangement, and are either arranged on the walls and ceiling, or built in a special chamber as dry or moist air coolers, according to the quality of the chocolate to be cooled, or the use for which it is destined. Whilst in the former case cooling is effected directly in the rooms, in the latter the air of the cooling room is conducted to the air coolers by means of ventilator, in order to be cooled and dried there, and then again introduced in the chamber.

II. The compressor, a gas suction and pressure pump, working both simply and complex, which draws the refrigerating medium out of the evaporator, compresses it, and forces it along to the condenser.

III. This condenser consists of a coil of wrought iron or copper pipes, which are enclosed in a barrel and are often described as the immersion condenser. There is another type, in which the pipes are united to one or more pipe-walls, introduced in a vessel which collects and drains off the condensations. In both cases the coil of pipes is played upon by a continual stream of water, in order that the gases which they contain may be condensed. The immersion condenser is generally employed when there is a plentiful supply of cheap water at hand, and the other in contrary cases. This latter condenser is provided with a separate liquid "after-cooler", constructed on analogy with the before mentioned immersion condenser. The counter current principle holds good in both types, and admits of a better using up of the cooling water. The liquid gas then passes on to the evaporator, where it is responsible for further refrigeration.

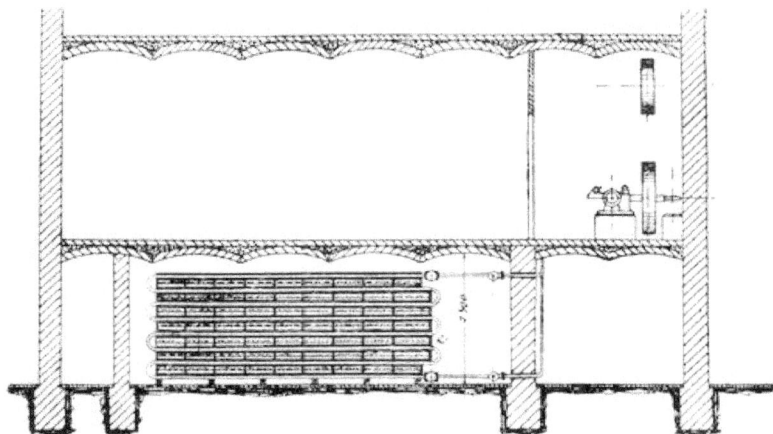

Fig. 59a.

The refrigerator also occurs in the form of a brine cooler. In this construction the evaporating pipes are likewise enclosed in a barrel, containing a high percentage of salt brine. In consequence of the refrigerating apparatus occurring on the interior of the pipes, the brine contained therein is cooled down to a very low temperature, pumped along to the cooling chambers, and after delivering its alloted refrigeration unit re-conducted to the cooling apparatus, where it is once more subjected to the same series of processes.

A well-known arrangement for such artificial refrigeration is that constructed by Wegelin & Hübner at Halle o. S., in which carbonic acid is employed, and it has been found well adapted for use in chocolate factories. The accompanying illustrations figures 59 a and 59 b represent an arrangement of that kind in which the cooling is effected on cooling trays

239

judiciously arranged.

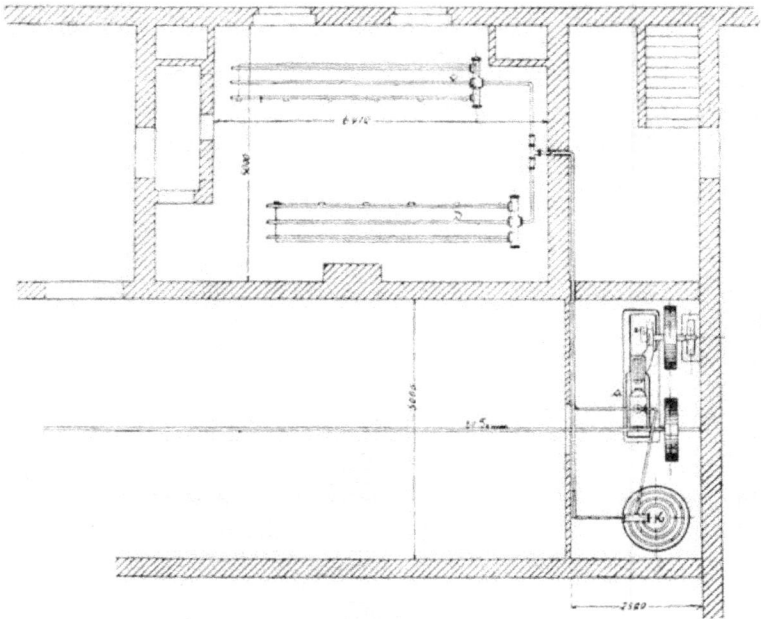

The refrigerating machine is constructed on the carbonic acid gas compression system; it consists of 1. the compressing pump **a**, 2. the condenser **b**, and 3. the system of pipes **c** and **d**, that constitute the refrigerator. The coil of pipes in the refrigerator is connected at one extremity with the compressing pump and at the other extremity with the condenser. Liquid carbonic acid passes from the condenser into the coil of pipes and is there evaporated. The heat necessary for that change is withdrawn, either directly or indirectly, from the cooling chamber and from the chocolate placed in it, until the desired reduction of temperature is brought about.

Fig. 60.

The compressing pump **a** is a peculiarly constructed suction and pressure pump, it draws out of the refrigerating pipes the vaporised carbonic acid by which they have been cooled and then subjects it to a pressure which helps to effect its reconversion into the liquid state.

The condenser **b** consists of a coil of pipes over which a current of cold water is kept flowing and the compressed carbonic acid vapour, passing from the compressor into these pipes, is there cooled and condensed by the surrounding water, so as to be transferred back to the refrigerator through a valve fitted to it for that purpose. The outer vessel of the condenser is constructed of cast-iron, in one piece with the compressor frame. These cooling arrangements are constructed either with or without mechanical ventilators. In figures 63 a and 63 b the compressing pump and condenser are represented as placed on the ground floor, while the refrigerator is situated in the cellar space lying beside them and at a lower level, in such a manner that both the systems of cooling pipes are not situated upon the roof of the cellar, but run along it at regular distances parallel to the side walls of the cellar. The compressor and condenser form one apparatus and the

241

former is driven by a steam engine.

In the cooling cellar, the refrigerator is generally fixed to the walls in such a way that the warm chocolate, taken into the cellar, can be at once placed upon the stages formed by the system of cooling pipes, and so there is some advantage in having the system of cooling pipes fitted along the roof of the cellar.

The machine which is diagrammed in fig. 60 possesses an hourly output of some 70000 calories, measured in salt water at -5 ° C. The compressor is driven directly by an electric motor, and a stirring apparatus is put in motion by the crankshaft of the compressor, the two being connected by an intermediate gearing.

Wegelin and Hübner put out cooling plants with salt water cooling, smaller and medium sized plants are on the contrary provided with so-called direct evaporation.

The diagram in fig. 61 shows an air-cooler as built of late by Esher, Wyss & Co. for chocolate cooling plants.

These air coolers are especially used for direct evaporation of carbonic acid gas. They consist of three groups of ribbed wrought-iron pipes, the whole constituting a system supported in a frame work of U-shaped and angular iron. The separate tubes are welded and bent together. The ribbed bodies are in themselves square shaped, and apart from the tube opening have a nozzle introduced in their centre, which pressed firmly against the press pipe effects a favourable transmission of heat in the case of large surface areas of the support, the more so as the tubes are square shaped.

Among the numerous advantages of this machine can be numbered the abolition of the refrigerator and brine pump, prompt and instantaneous refrigeration when the machine is started, and ease of control, as a flange connection occurs immediately in front of the machine.

Fig. 61.

A wrought iron trough is fitted up underneath the air-cooler to catch the water drops. Above, and to the left, the three systems of the air-cooler are connected by means of a catch.

In the foreground of the illustration is given a miniature of the ribbed tube system, which very clearly illustrates the arrangement of the separate ribs.

A ventilator not apparent on the diagram conducts air to the tubes in the cooling chamber, and these present a considerable cooling area, in addition to which, the air-stream taking a parallel direction, resistance to its passage is reduced to a minimum.

Another method of cooling[119], that is carried out in France consists in placing the moulds, containing cakes of chocolate, upon a travelling belt running horizontally through the whole length of the cooling chamber. The requisite reduction of temperature is effected by apparatus similar to that described above in Wegelin & Hübner's arrangement. The liquefied carbonic acid flows through a system of pipes fitted to the roof of the cooling chamber, producing by its vaporisation the necessary cooling and then it passes back to the refrigerating machine. Circulation of the air in the cooling chamber is provided for by a suitable ventilator under the pipes of the cooling system, gutters being fixed to carry away any water condensing upon their surface and prevent it from falling upon the chocolate. The travelling belt passes along so slowly that the moulds, containing chocolate, placed upon it at one end, take from ten to fifteen minutes in passing to the other end where they are taken off and carried to the packing rooms.

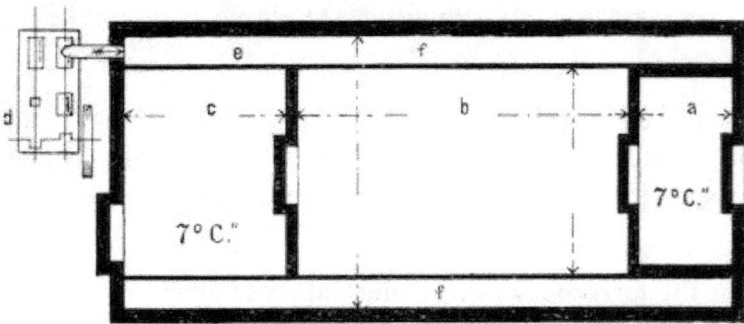

Fig. 62.

Another cooling arrangement that works very well is constructed by T. & W. Cole of the Park Road Iron Works, London E.; figure 62, represents a plan of this arrangement, which has the great advantage of providing for the exclusion of moisture from the cooling chamber. Refrigeration is effected, by means of Cole's Arctic-Patent Dry Cold Air machines, by compressing atmospheric air and then allowing it to expand, after being cooled by water and having moisture removed by suitable arrangements. The machine is of very solid construction; it works at a pressure of from 70 to 80 atmospheres and drives the dry cooled air through a system fitted in the cooling chamber where the chocolate is spread out, either on portable trucks or on a travelling belt, so that it remains in the chamber long enough to become perfectly cold. The system of cooling can be changed in various ways. The sudden removal of the cold chocolate into another chamber where the air is moist, would be attended with a deposition of water upon the goods. For that reason the goods are first transferred, for a short time, to a warm chamber (ante-room) where they acquire a temperature at which no deposition of moisture can take place. The chief advantage of this arrangement is that it furnishes dry cold air economically, both in summer and also in a moist climate. Cole guarantees that this machine will effect a refrigeration of 5 ° C.; according to the statement of Messrs. Negretti and Zambra the cooled air contains only 40 % of moisture. The cold air from one of these machines can be led, by a well insulated run of pipes, to any part of the factory and thus be made available for cooling purposes in different places.

The cooling plants hitherto described may be classified as "Space Coolers", because in each case a special compartment of the cooling chamber must be utilised. The increased prices of estate constitute no mean objection to such a system.

A critical valuation of these plants brings out a few undisguisable deficiencies. A large proportion of the cold is lost in the chamber itself, before it has been of any avail; and then again the rooms are generally insufficiently, sometimes even not at all, insulated from adjacent and warmer chambers, which once more involves raising of the low temperature essential in the process.[120] Detrimental also is the presence of the personnel, the illuminations, and many minor influences. It is evident that the larger the output required the larger must the cooling chamber be, involving corresponding economical waste.

With the recognition of these evils arose the problem of their abolition. The aim was to employ small chambers and avoid loss of cold air. It is now solved by a system already used in many and various industries, namely, cooling in closets. Larger or smaller closets may be employed, as required, and in consequence of their thorough insulation may even be introduced into the warmest rooms. Their principle is maximum efficiency with minimum occupation of space, and avoidance of loss of cold as far as possible. In consequence of this latter aim, the refrigerators in this case can be constructed on a smaller scale than those destined for an equal output of material, which are fixed up in cooling chambers; or they may be larger, which is yet more important, for the efficiency of the machine under consideration can be considerably increased by connecting it with one of the closets.

There are two sorts of cooling chambers, those which transport the moulds automatically, and those which contain layers where the moulds are placed one over another. Both types are cooled by the circulation of air, so effected, that cooled air currents are sucked up by a fan out of a tubular system fitted underneath a horizontal partition, and then forced along to the chambers above, where they are evenly distributed over the rows of sheet-iron, laden with moulds, or where they play upon the travelling belt which transports the moulds out of the cooling chambers. The air passes once more into the tube chamber on the opposite side, where it delivers up the warmth it has in the meantime acquired, to enter finally the same system of circulation as before. The general temperature of the closets is a mean between 8 ° C. and 10 ° C., and the cooling lasts from 20-40 minutes, according to the strength and size of the tablets. As the temperature never goes lower than 8 ° C., it is impossible for the tablets to become moist when exposed to the warmer outer atmosphere. Fig. 63a shows a Cooling Chamber built by J. M. Lehmann, which is adapted for a daily output of some 1000 kilos, and divided into compartments one above the other. The sections of this chamber, which in the illustration plainly shows the small amount of space required for its erection, are divided by vertical cross-partitions into four compartments, each of which is provided with a shelf or stand to take a charge of 10 cooling trays, and accessible by three spring-doors, thus giving as small apertures as possible and reducing the loss of cold when charging to a minimum. In addition to this, each compartment is fitted with a contrivance for regulating and, if necessary, completely cutting of the draught. The position of the system of pipes is shown by the two pipe-ends to which it is connected. On the opposite side, or front of the chamber, is the fan-drive, either a small electric motor, or shafting. The perforated cooling trays are visible through the open doors. The sides of the chamber consist of two layers of wood with thick slabs of cork between them. All chambers of this system, including those with automatic conveyance of the moulds, can be taken to pieces for transport, the single pieces afterwards only requiring to be fastened together again when erecting the chamber.—The chamber illustrated serves for cooling moulded chocolate. For pralinés and the like similar chambers are supplied, which are, however, smaller and lighter in construction.

Fig. 63b represents a cooling chamber with forced air circulation and automatic conveyance of the moulds, built by the same firm. This chamber, which, owing to the travelling belt conveying the moulds, is of considerable length, is nowhere connected with the outside air; the whole manipulation of the moulds is carried on through small adjustable openings at the points where the travelling band enters and leaves the chamber. The band consists of chains in links on to which wooden laths are screwed and its speed can be regulated to suit the size of the tables to be dealt with. The width of the belt and chamber can at any time be varied to suit the place of erection and correspond with the length.

Fig. 63 a.

Fig. 63 b.

Fig. 63 c.

249

As is to be seen from the illustration, this cooling chamber requires the minimum of attendance and thus complies with the principles lately adopted in all large factories, in which the tendency is to substitute as much as possible mechanical appliances for manual work. It will be seen from the preceding chapters that this tendency is especially marked in the moulding department, where automatic tempering, moulding and mould-filling plants and shaking tables have already been introduced. In order to utilise fully such automatic plants the last link in the chain only was wanting, namely, a suitable means of transferring the moulds from the shaking tables to the cooling chamber and through the latter to the demoulding and packing room. The purpose of the cooling chamber above described is to fill up this gap, and its proper place is thus ranged in among the automatic machinery described.

Thus it is that many modern factories have united the above machines to form a single working plant, as shown by Groundplan Fig. 63c.

V. Special Preparations.

a) Chocolate Lozenges and Pastilles.

These chiefly consist of cacao mass, sugar and spices. Formerly they were made by placing the semi-liquid chocolate material on a stone slab, furnished with a rim of uniform height which served to regulate the thickness of the goods manufactured, and then rolling out the mass as required. The lozenges were punched from the rolled-out layer by means of a cutter. After allowing the mass to cool, these lozenges were detached from the remaining portions, which were then rolled again and the same process repeated.

Pastilles, on the surface of which impressions of varying import, such as figures, names, firms etc. are required, may also be manufactured by placing the soft chocolate mass upon tin-plates in which depressions occur corresponding to the device desired. A roller is employed to make the material fit into the depressions, and superfluous chocolate is removed with a knife.

These impressions come out especially fine, when the pastille moulds are subjected to a shaking on the tables with which we are already acquainted.

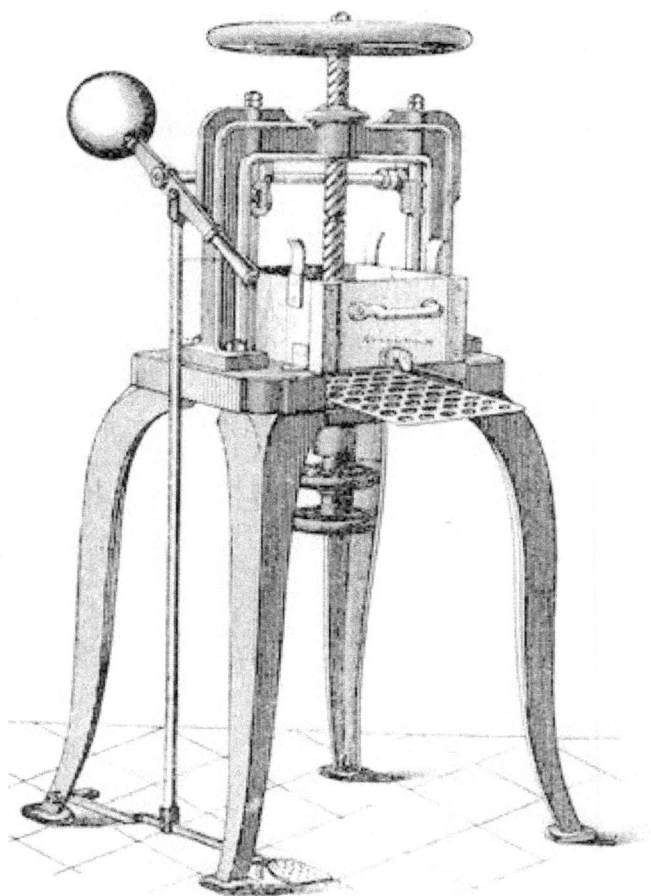

Fig. 64.

Fig. 65.

Yet these processes are becoming obsolete, and the chocolate slabs or plates are at the present time superseded by the two forms of apparatus constructed by A. Reiche, which we accordingly describe below.

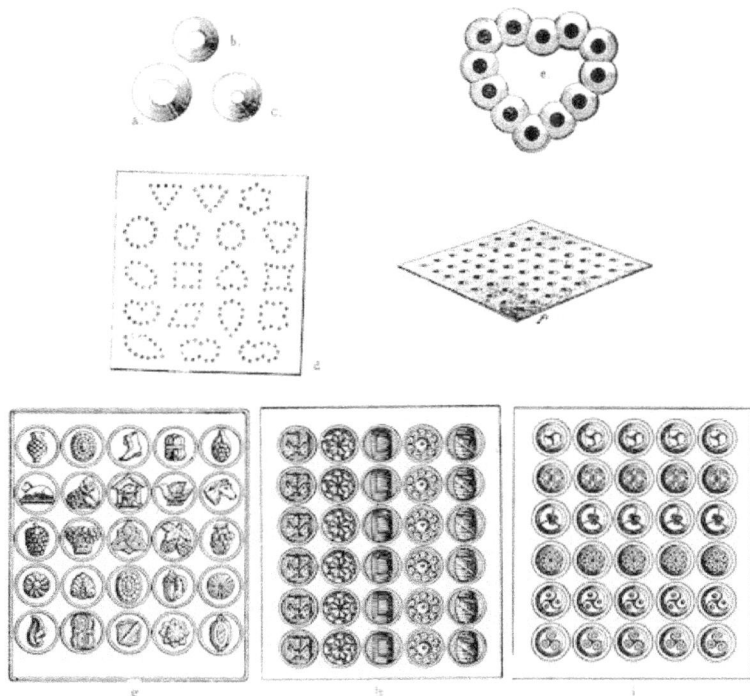

Fig. 66.

In the first of these simply constructed machines, fig. 64, the material oozes through perforations in a square sieve-like arrangement, at length issuing on the sheet-iron plate fitted underneath. The process is aided by repeated shaking, and when sufficient chocolate has penetrated to the plate, the box is raised on its hinge and chocolate mass left ready for further treatment. By gentle additional shaking, the still irregular heaps are rounded off to perfection; they are now cooled down and finally detached. The coating of the lozenges with coloured sugar grains is effected by passing them, together with the plate to which they still adhere, through a box containing sugar dust.

This machine is scarcely used now; in its place come the two constructions of A. Reiche, as already stated, the one being intended for solid material, and the other for semi-liquid chocolate mass.

Fig. 67.

By means of his pastille machine Nr. 14091, which we give
in Figs. 65 and 66, chocolate lozenges of the most diverse
size can be prepared very rapidly and to advantage. The

chocolate material, which in this case is solid, is pressed through perforations in a metal plate and otherwise treated as in previous cases.

In working with this machine, it must be previously and sufficiently warmed, then partially filled with chocolate material of a proper consistency (not more than 75 % of the total capacity may be utilised). It is highly important in the preparation of lozenges that the material should neither be too hard nor too plastic, but strike a just medium.

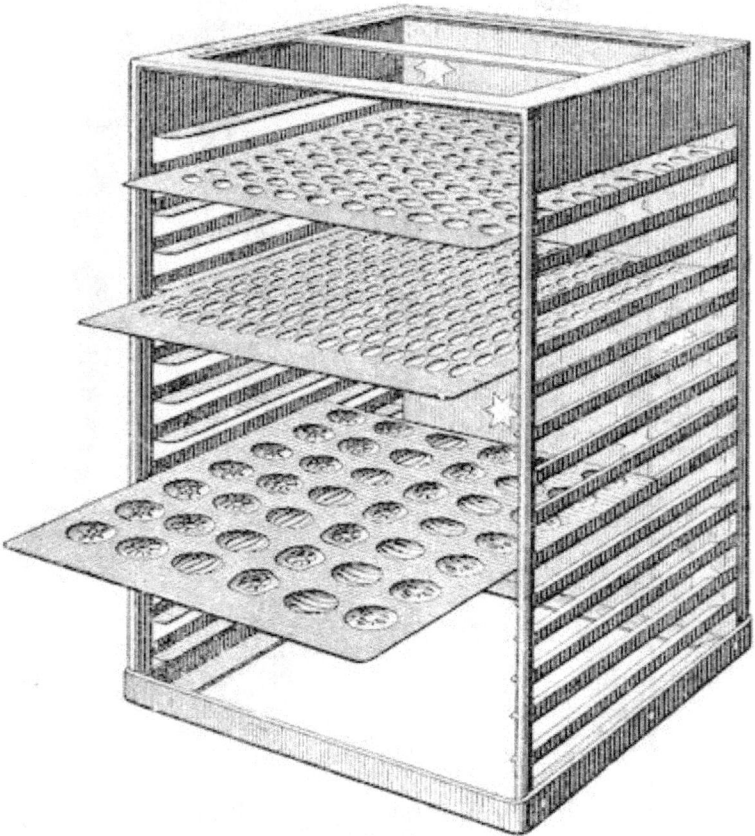

Fig. 68.

Before pressing down the plunger, worked by a screw, a metal plate is laid upon the chocolate to prevent contact with the plunger. By slight pressure, the chocolate mass is forced through the perforations, according to the required size of the lozenges, but the plunger must not be screwed down further. This will admit of the plate on which the lozenges rest being drawn out and another inserted.

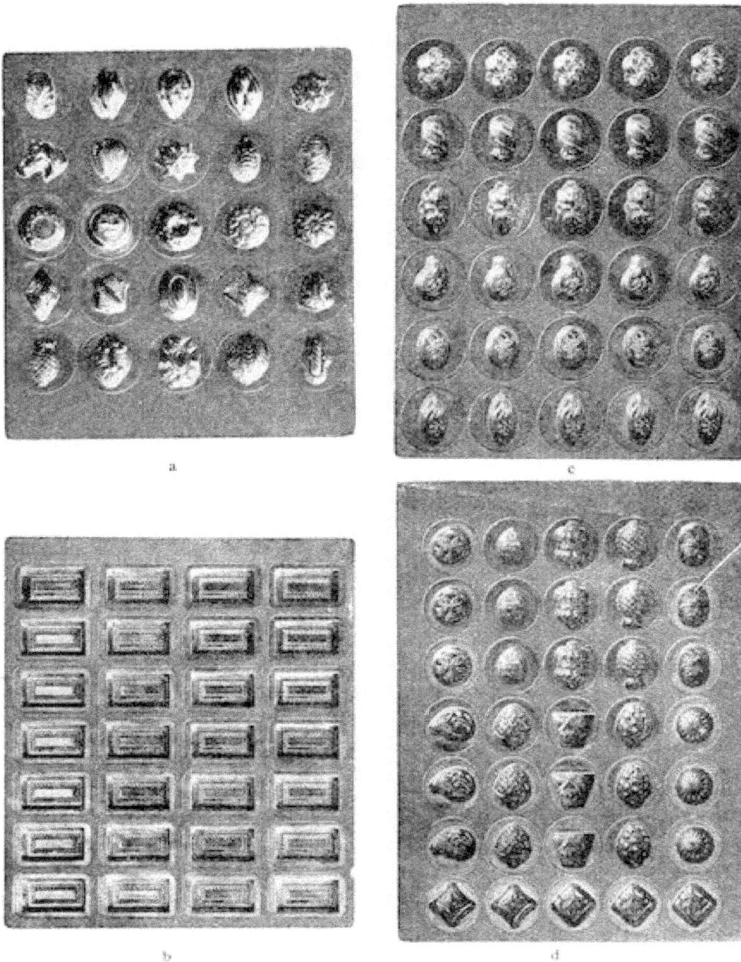

Fig. 69.

257

To this machine belong the usual perforated plates *f*, Fig. 66 of which there are three of different sizes for each machine, as shown by figures *a b c*, also the plates *d* used for making the perforated confections which find their way to the Christmas Tree. These plates are impressed with larger or smaller designs, and so make two different sizes of goods possible. A third plate is supplied for the manufacture of whole pieces (various varieties of chocolate croquette).

The machine works smoothly and noiselessly and delivers excellent products. If instead of the usual plain lozenges, such with the name of a firm or other device are desired, the corresponding impressions must be stamped out on the plate in which the chocolate is received after being forced through the perforations. See fig. 66, g, h, i.

Fig. 67 illustrates the pastille machine Nr. 14 178 for thin chocolate mass, constructed by A. Reiche (German Patent 227 200). It resembles the foregoing apparatus in principle and appearance, being only distinguished by a different aim, namely the treatment of thin material. Used in conjunction with the peculiar moulds also manufactured by the same firm (marked "Durabula"), even the deepest impressions can be effected with an enormous saving of time and material and in a most practical manner, as will be seen on comparing figs. 69 *a* to *d*.

In order to get the full value out of this machine, some little practice is necessary on the part of the workman in charge. But possessed of an average amount of skill, he can soon turn out with this apparatus ten times as much as can be made with the ordinary type of lozenge machine.

For a favourable accommodation of the different pastille plates, the hurdle diagrammed in fig. 68 (by A. Reiche) is quite excellent. It is manufactured out of one complete sheet of steel, is free from any suspicion of soldering, and entirely galvanised. It thus offers a strong guarantee as regards wear and tear. It may also be advantageously employed as a transporting device.

b) Coated chocolates, pralinés etc.

These delicacies are now held in high esteem, and of late the consumption of pralinés and cheaper forms with imitative contents has increased very considerably.

The designation praliné (properly pronounced prahlin) has been applied to sugar-coated almonds and is derived from the name of a cook in the employ of Marshal du Plessis, which was Pralins. This "chef" belonged to the age of Louis XIV. and was the first to make these sweetmeats. But now the term is applied to sweetmeats of various forms, soft fruit-sugar, marmalade, cream, nut-paste etc. respectively enveloped in chocolate. The special formulae employed in the preparation of different kinds of pralinés are comprised in the confectioner's art, and do not need to be dealt with here.

The substances themselves are called fondants. Formerly the sugar was boiled, placed upon a slab, and there manipulated with a spatula, an operation difficult to manage, indeed almost impossible in the last stages. In consequence of the increased demand for such preparations, machines were introduced several years ago whereby the operation is mechanically performed. Such a machine is shown in fig. 70.

Fig. 70.

The bed-plate as well as all the working parts of the machine are constructed of stout copper. The working parts admit of being raised or lowered by means of the hand-wheel above, and they remain fixed whilst the bed-plate turns and its underside is played upon by water. The machine is capable of working up pure fondant without any syrup addition, as well as that made up with syrup. The boiled sugar is poured on the bed-plate of the fondant machine, cooled down from 10-20 minutes according to the syrup content, and to such an extent that the machine can be set in motion, whilst the working parts are gradually lowered to the previously mentioned bed-plate. The sugar poured out is then cooled by means of the action of a ventilator fitted on a crossbeam, occurring in the middle of the wooden cooler, and working in conjunction with the ventilator, in consequence whereof a cooling current of air is brought to strike the hot sugar centrally.—When pure sugar is used, the fondant is finished within six minutes, but in the case of a syrup addition the time required is lengthened.

Fig. 71.

A quite recent type of fondant machine is given in fig. 71. It achieves its end by employing an air-current and a cylinder with screw, which is provided with water cooling apparatus. The *modus operandi* presents many and obvious advantages, chief among which is the possibility of conducting new material to the machine uninterruptedly, and further the preservation of the flavour of the chocolate worked up. The result is a production of first-class quality in respect to taste and flavour, which is quite ready to be passed on to the next processes.

Fig. 72.

The fondant is then diluted with colouring matter in boiling pans, and so prepared for subsequent treatment. The figures which have to be poured in are then transferred to gypsum moulds, lined with starch powder, and the fondant sugar is in its turn poured over these either by means of pans held in the hand or such as are machine-driven. Hand-pouring postulates a considerable amount of skill on the part of the man in charge, especially when even weights of the separate pieces are required. We annex an illustration of a motor-driven depositing machine (fig. 72).

The sugar is here introduced into receivers heated by means of a water-bath. The receiving boxes are moved under the outflow one after another, after having been dusted with powder and filled with chocolate, whilst the adjustment of the weight of each separate piece is effected by the operation of a very ingenious mechanism, even from 0-8 grammes.

Fig 73.

After a stay of several hours in the drying room, the molten figures are so hard that they can be raised out of the powder with the aid of a shovel. Fig. 73 shows such a machine, whilst Fig. 74 illustrates a machine where the work goes on unbrokenly, and from which the chocolate figures are removed with a shovel.

The sweetmeats are next dipped into liquified chocolate (covering stuff) to coat them with a layer of that material. The mass employed for this purpose must contain up to 15 % more butter than that used for ordinary chocolate, so that it may be kept soft long enough for continuous working.[121] This is performed in the machine fig. 75. On a bed-plate coming into contact with steam or cold water, as required, occur rake-like stirrers, and a small ventilator introduced above assists in cooling off the material. For the purpose of discharging, there is an outlet on the rim of the pan. For storage of the tempered coverings and also for occasional alleviations with cacao butter, a machine illustrated in fig. 76 is utilised.

Fig. 74.

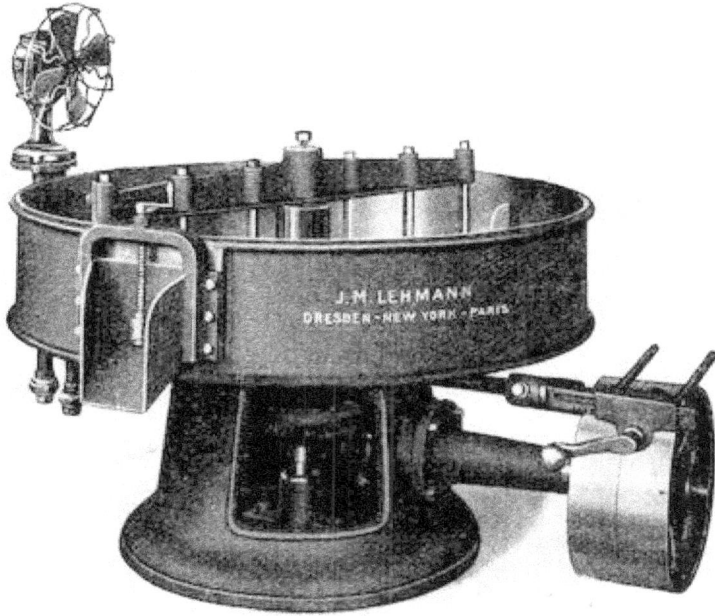

Fig. 75.

Fig. 76.

The dipping of pralinés for the purpose of coating them was formerly carried out by means of a fork, the nucleus masses being dropped into the coating material, taken out with a fork, and placed upon metal plates. Various kinds of ornamentation were designed by the same instrument. In the preparation of the higher priced coated fondants, a similar method of procedure is still in vogue, although such manipulation presupposes a high degree of skill on the part of the mechanics are at the machine. For articles of more general consumption, whether ornamented or not, machines have been introduced for the purpose by divers manufacturers, some of which function excellently. Two of that kind which in every way respond to the calls made on them are here described, but we shall not waste time and labour over the more complicated and expensive machines.

266

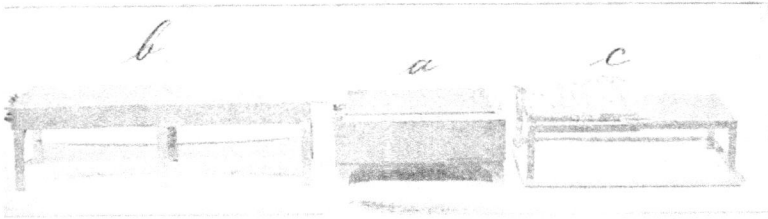

Fig. 77.

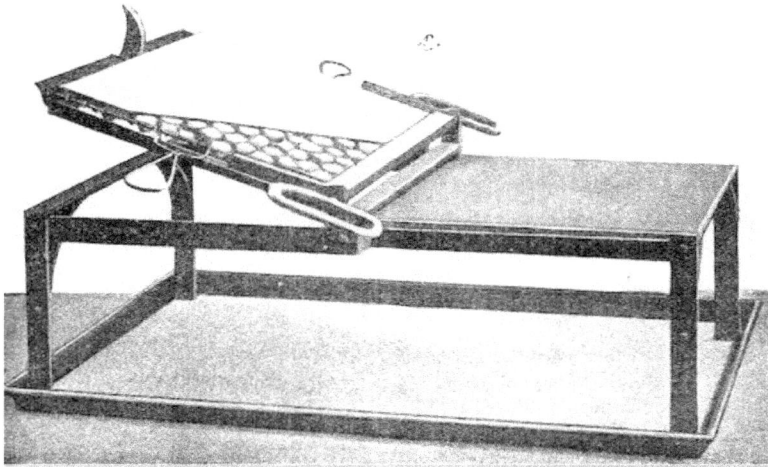

Fig. 78.

The first method of coating fondants, patented by A. Reiche of Dresden-Plauen, is not based on mechanical principles, but rather relies on a series of small appliances, represented in fig. 77. The jacketed casing *a*, fig. 77 contains water, and into it the pan containing coating material can be placed: that is kept in a liquid condition by heating the water in the jacket by spirit lamps or gas jets underneath. The adjoining vessel *b* is closed on all sides, filled with water, and also kept warm in the same manner; it serves for the preparatory warming of the objects to be coated, which are spread upon a wire network, and for that purpose two of these wire frames can be hung upon the hooks inside the box. The mass dropping from the wire frame is conveyed

267

into the covering box *a*, by means of a sheet of metal placed above it; *c* serves as an apparatus for turning, and we give it on a larger scale in fig. 78.

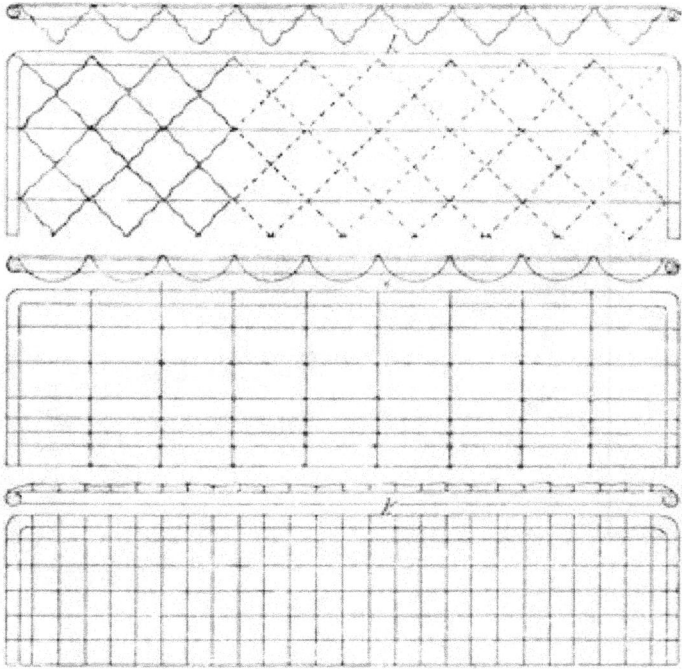

Fig. 79.

The tracings *h* and *i* in fig. 79 show the cross section and top view of the wire gratings, on which cylindrical and ball-shaped sugar goods are deposited. The other two kinds of grating are illustrated at *L* and *M* (fig. 80).

The size of the meshes of the sieve gratings depends on that of the centres to be coated.

The method of covering is as follows:

The centres for the pralinés etc. are placed in the cavities of the gratings, and, as soon as one of the gratings is full, the latter is covered up by the fine-meshed grating the half of the cross-section of which is shown in Fig. 79 and the full view in Fig. 80 (see K and N respectively), K representing the cover-grating.

Both gratings are held simultaneously by the operator at their handles and then dipped together in the liquid covering contained in the vessel a, Fig. 77, after which the superfluous covering mass is removed by knocking. The gratings are now deposited on the mechanism C, Figs. 77 or 78, as the case may be, the top sieve removed and a sheet of paper or a metal plate put in its place. It is then turned by hand to the opposite side, the grating with the impressions is removed and the cover centres are found lying in regular order, and at regular distances apart, on the metal plate. The object of the intervals between the covered centres is to prevent the running together of the latter.

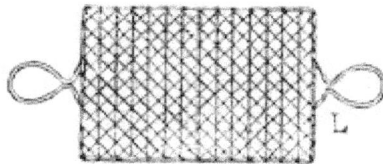

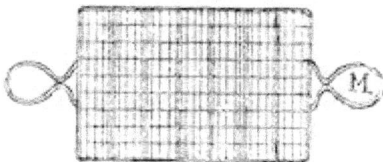

269

Fig. 80.

Beans and rings are only dipped up to the middle, and the process repeated with the other half of the centre after the first half has cooled. This ensures a pleasing, round appearance, and has further the advantage that the cover grating need not be put on during the operation. When dipping cylindrical or ball-shaped centres, the grating K which has first been removed on dipping, is at once transferred to the heater, to prevent it cooling and withdrawing too much warmth from the covering material at the next immersion.

The dipping of pralinés etc. is exceedingly easy if the new type of dipping machine is used, a full view of which is given in Fig. 81 and which has the highly appreciable advantage of simultaneously cooling the dipped centres. All the parts are, in the main, worked by hand, only the shaking and stirring contrivances and the cooling fan requiring to be driven by motor power. The middle piece carries the actual dipping apparatus, underneath which the tank holding the covering chocolate is fixed, while the lefthand sidepiece serves for feeding; as many as four operators can be engaged simultaneously at the latter, the work consisting of laying the centres in the gratings corresponding to the mouldings desired. The construction of these gratings is, in the main, similar to the stamped trays of Anton Reiche, but they are not provided with handles and are despatched along the guide-rails by hand. The filled grating is then placed in a frame, which is dipped by means of a winch into the liquid chocolate. The top grating on the dipping frame is adjustable, and the object of this grating is to keep the centres down, as without this arrangement some of the centres might rise to the surface of the covering. The top grating is, before commencing to dip, pushed over the filled grating with the centres and is thus immersed with them. The frame having been removed, the shaker is put in action to remove the superfluous material from both the gratings and the centres. The grating is drawn out after use from below the top grating and transferred to a book-shaped ejector, on one side of which is a metal sheet covered with paper. The whole of the centres are then discharged on to the sheet, by reversing the two flap-sections.

The sheet containing the covered centres is then transferred to the cooling apparatus at the right, in which it is gradually lowered on a "paternoster" apparatus by turning round a handle. It is then conducted to the left by an endless band, and finally discharged in a cooled state by the machine. The ventilator should be supplied with air from the cellars and is arranged to blow it out in the opposite direction to the goods in the cooling apparatus.

Fig. 81.

The shape of the design-gratings is reproduced in high relief on the goods, and it will therefore be readily understood that further designs or fancy shapes can be made on the gratings. For the production of semi-dipped goods or such as are dipped round and remain uncovered at the bottom, a device is attached to the striking gear which renders it possible to regulate the depth of each immersion at will. The tank containing the covering material is surrounded by a water-jacket, which is heated by steam. The heat of the water is indicated by a thermometer. The receptacle containing the covering can easily be drawn out towards the front. In addition to this, the whole of the outside of the machine, which also constitutes a complete water-jacket, is heated by steam, and finally the ejector. The gratings containing the impressions are taken out of the ejector after use and transferred to the feeding side to be used again, so that, at the very most, four gratings are required for each design.

The daily output of the machine is 300-600 kilos, and the size of the gratings 280 by 400 millimetres, the output naturally depending on whether the machine is operated by two, three or more persons.

B. The Manufacture of Cocoa Powder and "Soluble" Cocoa.

a. The various methods of disintegrating or opening up the tissues of cacao.

The comparatively high fat content of pure cacao, which would deter certain persons, especially those suffering from stomach disorders, from taking it, has given rise to the now extensive demand for a cacao preparation containing a less amount of fat and the constituents of which are capable of being easily assimilated in the human organism. At the same time the desire to obtain a cacao preparation easily capable of complete and uniform suspension in milk or water may have played its part, as this quality, in consequence of which the preparation can rapidly and without difficulty be rendered ready for consumption, is obviously a great advantage. The best way to obtain this appeared to be the pulverisation of the cacao, which, when reduced to a powder, more readily satisfies the above conditions. As, however, it was not possible to pulverise cacao which still contained its full amount of natural fat, it became necessary to devote attention to the operation of extracting the cacao butter. It is many years since the first appearance of certain preparations in Germany which went under the name of "Cacogna", and which had been deprived of their fat to the extent of 20-25 %. This problem, however, was recognised and attempts and all manner of experiments made to solve it at a much earlier period in Holland. The founder of the well-known Dutch firm of J. C. van Houten & Sons in Weesp, Mr. C. J. van Houten, was the first to attempt the expression of the fat from cacao (1828) and to treat it with chemical agents with a view to opening up or bringing about the disintegration of the tissues, in order to render the cacao a fit and welcome article of food, not only for healthy persons, but also for invalids and convalescent persons.

It was not until the Dutch cocoa thus manufactured had been introduced into England and Germany, where, as well as in Holland, it became very popular, that manufacturers in Germany and Switzerland began to devote their attention to the treatment with chemical agents. The consumption of so-called "soluble" cocoa has increased to such an extent of late years that it is now almost as large as that of chocolate goods.

The term "soluble", as now generally applied to cocoa powders, is undoubtedly a misnomer, inasmuch as such preparations are practically not soluble at all. We have therefore termed cocoa for drinking purposes in this book "disintegrated" cacao, as the processes described in the following pages only render the elements of cacao, as, for instance, the cellulose, capable of suspension in liquids. It would be quite impossible to render cacao, by any special treatment, soluble in the real sense of the term, as is the case with salt or sugar. It will thus be readily understood that the expression "disintegrated" is correcter and more logical than the term "soluble The degree to which disintegration has been carried, i. e. the efficiency of the opening-up processes adopted, is marked by the absence of any sediment worth speaking of in the beverage prepared with boiling water, even after it has been left standing some time. The greater the power of suspension of the preparation, the less particles of cacao will settle to the bottom, and the higher the beverage will be esteemed.

The disintegrating agents are, in practice, applied either to the raw or roasted, but otherwise untreated beans, or to the more or less defatted cacao, as follows:

a) by treating the cacao with hot water, without or under pressure;

b) by treatment with alkalis, such as carbonate of kali or sodium, carbonate of magnesia (Dutch method), spirits of ammonia (sal-ammoniac) and carbonate of ammonia (German method).

The chemical and physical effects brought about by these agents consist chiefly in the swelling or steeping of the cellulose by the action of the alkalis, as a consequence of which they sink less rapidly in liquids than would be the case with untreated cacao. A further effect is the partial neutralisation of the acids present, besides which the cacao-red or pigment is also attacked, a result which may be regarded as less desirable, as the cacao-red is the secreter of the aroma, which naturally suffers with it. If the cacao is treated with steam or hot water, the starch is apt to gelatinise, and the acids to begin to ferment.

As the treatment with steam, for the reasons given above, is nowadays rarely practised, we will at once proceed to consider the method of disintegrating cacao most in use. Modifications of the methods of manufacture bearing on this point will be dealt with in their place under the corresponding heading later in this book.

b. Methods of Disintegration.

1. Preliminary Treatment of the Beans.

The method of manufacture of disintegrated cocoa comprises the following operations:

a) The cleaning and sorting of the raw bean;

b) Roasting;

c) Shelling, breaking and grinding;

d) Treatment with alkalis or water;

e) Expression of the fat or cacao butter;

f) Pulverising.

The order of the above processes is subject under certain conditions to various modifications arising from the fact that the alkalis are applied at various stages in the course of manufacture, i. e.:

I. before roasting;

II. during roasting;

III. after roasting,

and further

a) before pressing;

b) after pressing (treatment of the defatted beans).

The cleaning and sorting of the raw beans, or, in short, the complete treatment to which the raw cacao is subjected (a to c) is in all methods effected by the same machines, a description of which has been given on pages [Transcriber's Note: Rest of line missing]

Some manufacturers proceed at once to treat the cacao with alkali on completion of the above operations.

C. Stähle[122] effects the disintegration of cacao by subjecting the beans to the chemical action of a mixture of ammonia and steam, at a temperature not exceeding 100 Deg. C. The next process (roasting) is then supposed to draw out the ammonia introduced into the material, which, being volatile, easily escapes, and enables the flavour to develop.

Pieper[123] moistens the raw beans with water, to which alkali has been added, and this has the effect of neutralising the acids present in the bean; afterwards the beans are fermented, dried and roasted. The fermentation is described as rendering the particles of albumin or protein bodies easily digestible and further imparts to the beans a fine, reddish brown colour. This process is therefore nothing but an after-fermentation of the cacao under the influence of alkalis. From a scientific point of view, the process does not possess the advantages which Pieper claims for it, with the exception of the really evident improvement in colour. This effect can, however, be obtained equally well by suitable treatment with water alone.

G. Wendt[124] has patented a method of improving the colour and facilitating the disintegration of cacao, in which the beans are treated, before roasting, with lime water and milk of lime (lime solutions) and further washed with the solution during roasting.

We now turn to the methods of disintegration by means of fixed alkalis (carbonate of magnesia, potash and sodium) first employed by the Dutch, concerning which the following description will be useful.

The cleaned beans are first very superficially roasted, to facilitate winnowing, and the cacao thus treated (half roasted cacao) broken as small as possible, which is an equally important factor in the shelling and winnowing processes. It should be observed here that the less the cacao has been roasted, the finer it should be broken. The material is then impregnated by one of the above-mentioned alkaline solutions, which is sprayed on to the beans. The chief agent employed is potash (carbonate of potassium) in the proportion of 1½-2 (3 at the outside) parts of potash to 20-30 parts of water, for every hundred parts of the defatted material to be treated. Some manufacturers use sodium or a mixture of sodium and carbonate of magnesia in place of the potash. As soon as the cacao has been uniformly impregnated by the alkaline solution, the roasting process should be completed. Still more care should be devoted to the roasting of cacao for pulverising than is required in the case of eating chocolates, as taste and smell play a more important part and the point of complete roasting is not so easily recognised. The cacao being roughly broken and the shells removed, the second roasting process must of course be conducted over a low fire. The most suitable machines for this purpose are the large roasting machines illustrated on page 93, Fig. 14, as in these machines there is little possibility of over-roasting, even when dealing with large quantities and the machine is intensely heated; another advantage is the easy accessibility of the roasting drum, which can be immediately exposed by removing the front cover, for cleaning; cleaning is very necessary in roasting machines. Broken and moistened cacao chars much more readily than raw beans which have not been deprived of their shells. If it is not possible to thoroughly clean the interior of the roasting drum, as is often the case with spherical roasters, the particles of cacao remaining in the drum continually undergo re-roasting, finally falling in a completely charred state into the cacao, thereby greatly prejudicing its taste.

If necessary, the cacao can now be passed through the breaking machine again, from which it is transferred to the triple cacao mill, which provides for fine grinding. The material is then deposited in heated pans (see page 117, Fig. 27) where it remains until ready for the next process, the expression of the fat. The object of the fine grinding in the mill is to render the cacao on being ground again after the defatting process, easily capable of being sifted, and to

278

obtain a preparation which, on being mixed with hot water, leaves as little sediment as possible.

2. Expression of the Fat.

Hydraulic presses are nowadays exclusively used for this most important operation in the manufacture of "soluble" cocoa. The methods of pressing have, in common with the other operations in the course of manufacture, undergone considerable modification and improvement.

According to Macquer (see Mitscherlich, S. 58) the butter was extracted during the last century by pulverising the seeds, boiling them in water and cleansing the fat, which, on cooling, congealed on the surface of the water, by re-melting. According to Desprez (see Mitscherlich, S. 58), burned, shelled and finely pulverised beans were spread to a height of 12-15 inches on coarse linen or canvas, which was spanned across a vessel containing boiling water, to expose the fine powder thoroughly to the action of the hot vapour. The powder was then pressed, in linen bags, between two tin plates, whereby some 50% of pure cacao butter was obtained. At a later period the heatable hydraulic pot-presses came into use. The mass had, however, to be introduced into these pots tied up in a cloth or sack, to facilitate which it was previously treated with water, forming a thickish syrup very convenient for pressing. All these methods, however, were attended with the great disadvantage that the cocoa, after being stored some time, acquired a grey colour, or became mouldy. To avoid these undesirable results presses were constructed which rendered it possible to liquefy the mass without any further treatment in the receptacle in which the pressing was conducted. Such a press, likewise acting hydraulically, is shown in Fig. 82 on the opposite page.

This machine exerts a total pressure of 320000 kilogrammes and works with 400 atmospheres. The construction of the machine is similar to that of the well-known types of presses used by oil manufacturers for the preparation of vegetable oils. When pressing, however, the pots containing the cacao must be rendered water-tight both at the top and bottom, to prevent the liquid cacao from escaping, while such provision is not necessary in the case of the oil presses. The stopping up of the press-pots is effected by means of a side-handle, and arrangements are provided for heating the pots both from above and below. The machine illustrated has 4 pots, arranged one above the other, which can be drawn out on guide-rails towards the front of the machine. During pressing, they close telescopically with the piston arranged underneath each pot. The pump which supplies the water for the hydraulic pressure, works perfectly automatically, increases the pressure according to the quantity of fat which has run off and keeps the pressure at its maximum or at any degree required. With these presses it is possible to extract, without difficulty, 85 % and even more of the total fat of the cacao bean. If pressing is carried on at too high a temperature, a pale, whitish grey butter is the result. If, however, a little attention is paid by the operator at the press, the butter obtained is usually perfectly clear, as it is first conducted through a horse-hair pad covered with linen, or a camel-hair cushion 15 millimetres in thickness. Sufficient attention is not always paid to the operation of pressing, so that it often happens that some of the cacao escapes with the butter, which is especially the case if the pressure has been increased too rapidly at the beginning. If the butter is extracted for use in the factory itself, the escape of the cacao with it is of no serious consequence; if, however, the butter is intended for sale for commercial purposes, its appearance is a most important factor, wherefore it is advisable to filter the impure fat immediately after pressing. It is true that, in most factories, the butter is in such cases merely remelted to allow the impurities to settle to the bottom, this part being then submitted again to the same treatment, while the rest of the butter is disposed of on the market. If filtering is necessary, the butter filter should be used, which, first constructed in Holland, has been in use for a long time there. The principle of these filters is to pass the butter through hanging tubes made of a filtering material similar to flannel.

Fig. 82.

The firm of Volkmar Hänig & Co. constructs special cocoa butter filters which can be obtained through the firm of J. M. Lehmann. Figs. 83 and 83a show this type of filter (cross and vertical section), the manner of working with it being the following:

As soon as the butter has passed through the hair sieve in the upper part of the apparatus, which removes larger objects such as pieces of wood etc., it enters the hanging filter tubes, which, to facilitate cleaning, are interchangeable. The filter butter accumulates in the large space provided for the purpose and is withdrawn through a tap. An observing glass is attached to the apparatus for the purpose of watching the height of the butter, and the whole filter is water-jacketed, the water being heated by a steam coil fixed in the bottom of the apparatus. A thermometer is fixed to the side of the filter, for regulating the temperature.

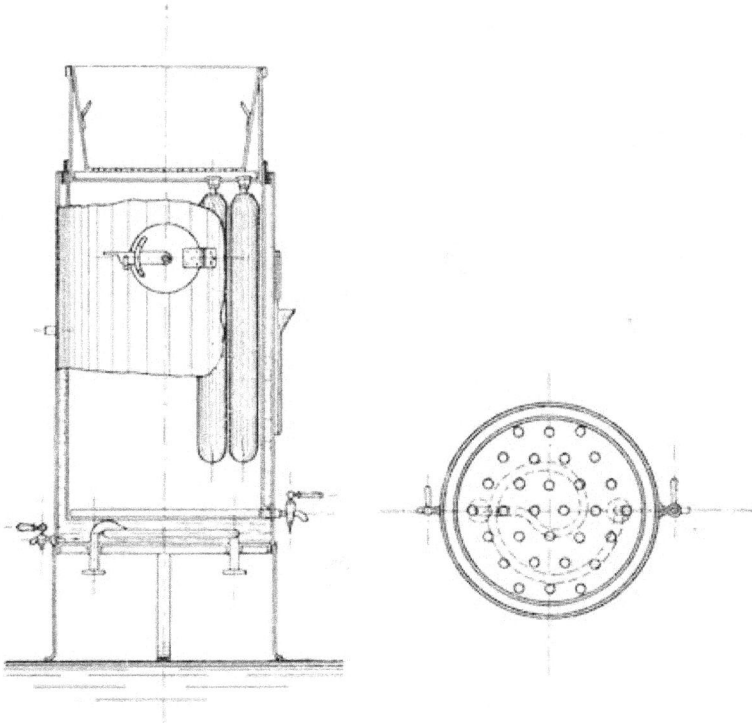

Figs. 83 and 83 a.

The degree to which cocoa powders should be defatted is

an important question which, some years ago, formed the subject of much controversy. The relation between the percentage of fat contained in the original cacao kernel, the expressed butter and the defatted cacao mass is shown in tables 19 and 20.

The taste of defatted cacao is, as is well known, all the better for being defatted to a low degree, and it is this which constitutes the great advantage of cocoa prepared according to the Dutch method, the remaining cacao content of which is some 24-33 percent, so that, taking 50 percent as the average quantity of fat contained in the cacao, only about 34-52 percent of the whole is removed from the mass.

Table 19.
Percentage of butter to be extracted.

Percentage of fat to remain in the finished cocoa powder	Fat content of kernel						
	50%	51%	52%	53%	54%	55%	56(
	Weight of butter to be expressed (in proportion to the whole mass)						
33%	25·4	26·9	28·4	29·8	31·3	32·8	34·4
32%	26·5	27·8	29	30·9	32·4	33·3	35·3
31%	27·5	29	30·4	31·9	33·3	34·8	36·2
30%	28·6	30	31·4	32·9	34·3	35·7	37·1
Fatty Cacao 29%	29·6	31	32·4	33·8	35·2	36·6	38
28%	30·6	31·9	33·3	34·7	36·2	37·5	38·9
27%	31·5	32·9	34·2	35·6	37	38·4	39·7
26%	32·4	33·8	35·1	36·5	37·8	39·2	40·5
25%	33·3	34·7	36	37·3	38·7	40	41·3
24%	34·2	35·5	36·9	38·2	39·5	40·8	42·1

	23%	35·1	36·4	37·7	39	40·3	41·6	42·9
	22%	35·9	37·2	38·5	39·8	41	42·3	43·6
Non-Fatty	21%	36·7	38	39·2	40·5	41·8	43	44·3
Cacao	20%	37·5	38·8	40	41·3	42·5	43·8	45
	19%	38·3	39	40·7	42	43·2	44·5	45·7
	18%	39	40·2	41·5	42·7	43·9	45·1	46·3
	17%	39·7	41	42·2	43·4	44·6	45·8	47
Diminution	(16%)	(40·4)	(41·7)	(42·9)	(44)	(45·2)	(46·4)	(47·
in value K.	(15%)	(41·1)	(42·4)	(43·5)	(44·7)	(45·9)	(47·1)	(48·

Table 20.

Percentage of butter remaining in the finished cocoa powder.

Weight of butter to be expressed, in proportion to the whole mass	Fat content of kernel						
	50%	51%	52%	53%	54%	55%	56%
Fatty Cacao 30%	28·6	30	31·4	32·9	34·3	35·7	37·1
31%	27·5	29	30·4	31·9	33·3	34·8	36·2
32%	26·5	27·9	29·4	30·9	32·3	33·8	35·3
33%	25·4	26·9	28·4	29·9	31·3	32·8	34·3
34%	24·2	25·8	27·3	28·8	30·3	31·8	33·3
35%	23·1	24·6	26·2	27·7	29·2	30·8	32·3
36%	21·9	23·4	25	26·6	28·1	29·7	31·3
Non-fatty 37%	20·6	22·2	23·8	25·4	27	28·6	30·2
Cacao 38%	19·4	21	22·6	24·2	25·8	27·4	29
39%	18	19·7	21·3	23	24·6	26·2	27·9
40%	16·7	18·3	20	21·7	23·3	25	26·7

	41%	(15·3)	16·9	18·6	20·3	22	23·7	25·4
	42%	(13·8)	(15·5)	17·2	19	20·7	22·4	24·1
	43%	(12·3)	(14)	(15·8)	17·5	19·3	21·1	22·8
Diminution in value K.	44%	(10·7)	(12·5)	(14·3)	(16)	17·9	19·6	21·4
	45%	—	(10·9)	(12·7)	(14·5)	16·4	18·2	20
	46%	—	—	(11·1)	(13)	(14·8)	16·7	18·5
	47%	—	—	—	(11·3)	(13·2)	(15·1)	17
	48%	—	—	—	—	(11·5)	(13·5)	(15·4

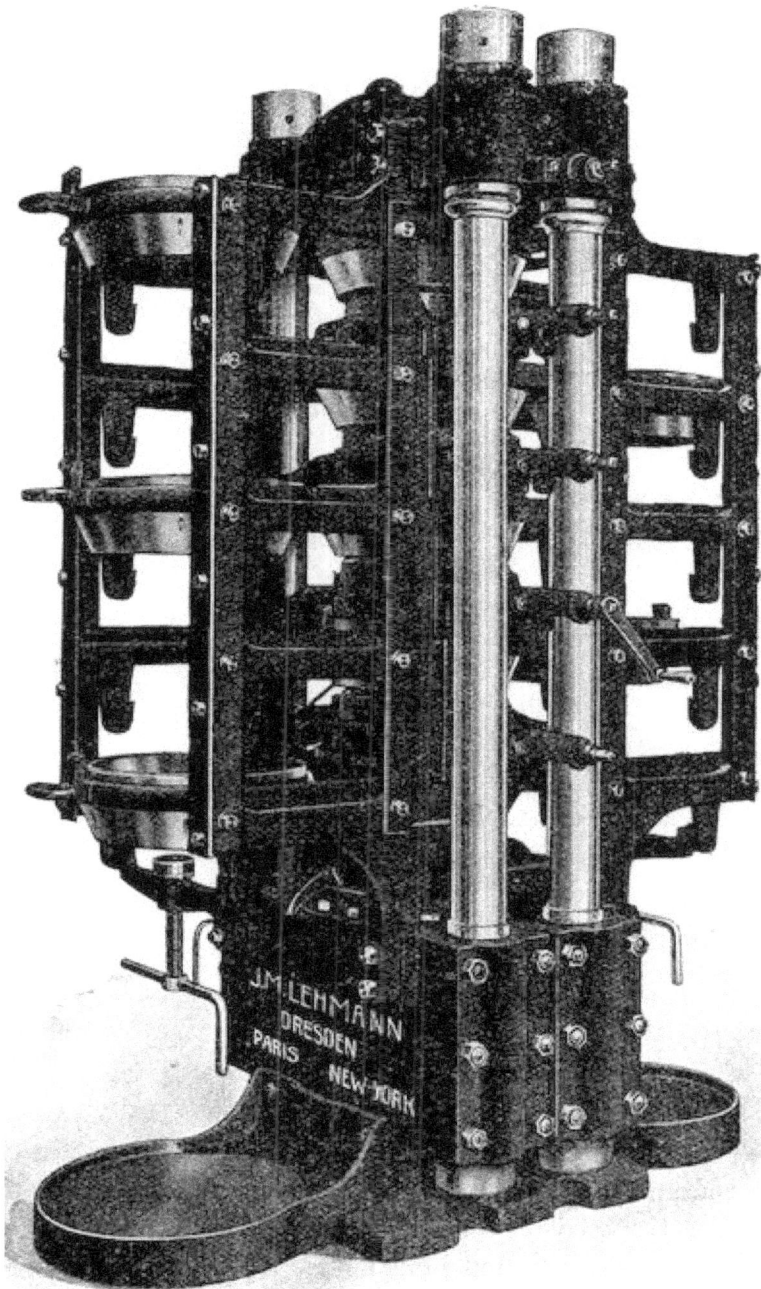

Fig. 84.

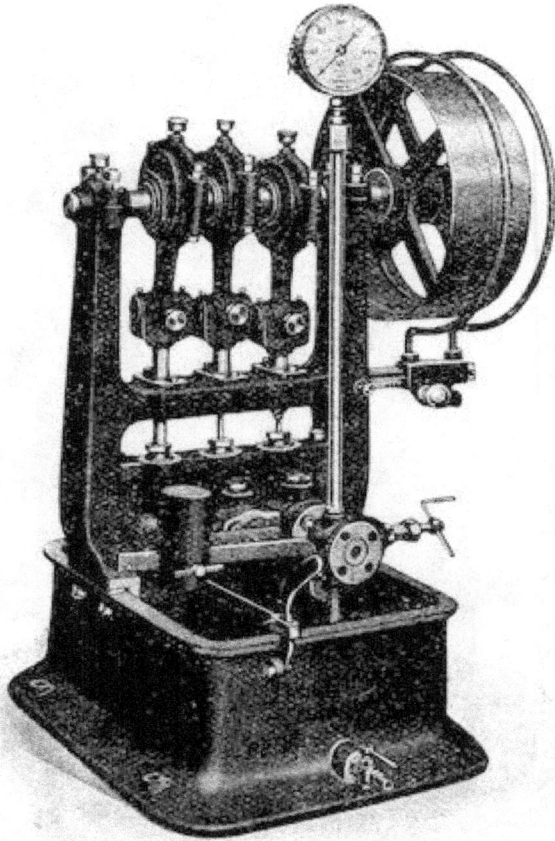

Fig. 84a.

If the expression of the butter is carried to a further
degree, the cacao will certainly become more easily capable of
suspension in liquids[125], but such treatment is detrimental
to its flavour[126], which is apt to become woody or bitter.
The statement, made by certain manufacturers and would-
be connoisseurs, that the bitter taste peculiar to the acid
produced in cacao during fermentation is the real aroma of
the cacao, is undoubtedly erroneous. It could, in the same
way, be said of tea and its acids, the bitterer, the better;

which would of course end in the destruction of the true flavour. Equally erroneous is the theory that bitter cacao is more consistent. Such cacao must, previous to consumption, either be more sweetened than usual or, if the same quantity of sugar is put in, less of the beverage can be taken. When, for instance, very thin coffee is made, the beans, on colouring an abnormally large quantity of water, are said to be stronger, i. e. to yield more. The consistency of all such beverages is, however, only a matter of taste, and it would therefore be useless to discuss the subject in detail; some persons prefer strong tea, which has been brewed a quarter of an hour, others simply pour boiling water over the tea leaves and then drink the beverage immediately. It may, however, safely be taken that the highest amount of butter which can be expressed from cacao without prejudicing the flavour of the finished powder is 66 percent of the total fat content. Manufacturers nowadays try as a rule to express as much butter as possible, as the butter has a high price on the market, and this tendency naturally has the effect of lowering the quality of the cocoa. We thus come across cocoa powders containing only 20, 17, 15 percent of fat and even less. Of course nothing can be said against the production of such cocoas, provided they are sold at a lower price than cocoas more rich in fat and the public are aware that they are purchasing a non-fatty preparation, besides which the expression of so high a percentage of the fat alone rendered cocoa a fit regular beverage for certain classes of invalids and persons suffering from disorders of the stomach. The only serious drawback in this case is the great variability of the fat content, which fluctuates between 13 and 35 percent. Such fluctuations are absolutely impossible in the case of any other article of food which is manufactured and sold wholesale, or, at any rate, buyers know in all such cases exactly what they are purchasing; this is a point to which serious attention must be called. It is very much to be regretted that the Association of German Chocolate Makers[127] has declined to follow up this matter, while the Union of German Food Chemists, after considerable controversy, advocated a distinct legal classification of non-fatty cocoa powders containing up to 20 per cent. of fat.[128] We would prefer the Dutch preparations, which have remained the same up to the present day, so-called fatty cocoas containing more than 25 percent of fat, to be classified specially and those preparations which contain less than this percentage of fat to be termed "highly defatted" or "dry" cocoas, the names applied to both kinds being of little importance as long as the public has the means of clearly recognising the distinction (see tables 19 and 20). Some 17 percent must be taken as the minimum permissible butter value, which would mean the expression of about 80 percent of the total fat content, or two-thirds of the cacao mass itself; cocoa powders with only 15 per cent. or less of butter are to be regarded as inferior in quality and should not be produced. Unfortunately, however, these suggested limits are, at any rate for the present, not likely to be realised.

Fig. 84b.

Fig. 85a.

The pressure obtained by means of the pressing devices above described is naturally not sufficient for the production of such highly defatted cocoas. Stronger presses are therefore necessary, one of which, a very powerful apparatus, is shown in Fig. 84.

This machine, at the present time the most powerful cocoa butter press in the world, brings a pressure of over one million kilos to bear on the cacao mass, working with 400 atmospheres, and thus renders it possible to express as much as 90 percent of the total fat content of the bean. The construction of this press is exactly the same as that shown in Fig. 82, the pump Fig. 84a having, however, three pistons or plungers instead of one; it works, like the other machine, automatically, i. e., after the large quantity of water required at the commencement has been fed into the press, the large plunger is put out of gear at a pressure of 5 atmospheres; the two smaller pistons are then put into action together, and produce the enormous pressure of 400 atmospheres.

Fig. 85b.

After defatting, the expressed cacao cakes are allowed to cool down, for which purpose they are transferred to flat trays or other suitable receptacles, and pulverising and sifting the powder thus obtained commenced.

3. Pulverising and Sifting the Defatted

Cacao.

There are several methods of proceeding with these operations, such as treating the expressed cacao in the melangeur already described in an earlier part of this book (cf. 30-32 figs.) or passing it through the centrifugal sifting machine (cf. 84b and 88 figs.) with which we are now acquainted. At a time when the melangeur was to a certain extent the universal machine of the manufacturer, it was almost exclusively utilised for pulverisation, that up-to-date division of labour whereby this machine is limited to mixing (and very properly so limited, as its name implies) and the preparation of cacao powders on the contrary assigned to more efficient constructions having then not as yet been adopted. We annex a description of one or two specially constructed arrangements for the pulverisation and sifting of cacao, as manufactured by J. M. Lehmann and already repeatedly tested.

Fig. 86.

First there is the cacao cake crusher (figs. 85a and 85b), which reduces the pressed cake into rather large pieces about the size of a walnut, previous to its being pulverised either in the melangeur, centrifugal sifting machine or some similar apparatus. It has been furnished with one (fig. 85a) and in some cases even with two (fig. 85b) pair of toothed or cogged rollers, and the cacao in this latter type of construction is crushed as small as a pea, which reduction, although it is by no means essential, considerably relieves the strain on the pulverising machine and is also in some sort a protection against unnecessary waste of material.

Fig. 87.

Then again, there is the so-called pulveriser shown in fig. 86. This is in principle an edge-mill with revolving bed-stone and runners, both made of granite. The coarsely broken press cakes are fed into the mill through a hopper provided with a slide, and are reduced to a loose powder of firmly fixed colour, escape of dust being prevented by the hood fitted to the mill. By turning a crank, a lateral sliding door is opened, and an arrangement inside is set in motion, by which the ground cacao is turned out of the mill. The pressure of the runners can be diminished and even completely nullified.

For cacao that has been thoroughly defatted ("dry" cocoa), the hardness of which demands a more efficient treatment than is possible in these machines, they being only calculated to press or at the most exert a rubbing effect, there are the crushers proper, called mills built in pulverising plants for dry cocoas as illustrated on fig. 87.

The pressed cacao, already broken up to some extent in a preliminary crusher (cf. figs. 85a and b), is systematically conducted through the mill by an elevator provided with hopper and feeding apparatus. On the interior of the machine, which is completely plated with steel-plates, there is a cross-arm as on a windmill, which passes through a large number of revolutions per minute. Chiefly owing to its thrashing effect, the cacao in the mill is fine ground, without any rubbing or exertion of pressure as in the melangeur and other machines. The outer part of the frame consists of a grating with various widths of hole, which can be readily changed. The whole of the powder which has attained a certain degree of fineness falls through these meshes and is so despatched from the machine at once, an additional advantage when comparing this mill with the melangeur, in which all the powder, even that sufficiently ground, must remain till the final discharging, much to the detriment of its flavour and aroma.

Fig. 88.

297

The powdered cacao next succeeds to the sifting operations, after it has first cooled a little, and for these the centrifugal sifting machines are used in the main. Special care must be taken that such apparatus as is used is not too diminutive to deal with the quantities of cacao introduced, as this is extremely injurious to the machine. It is further to be noted that no type of sifter whatever can yield good results if it has not been especially constructed for dry cacaos.

We have before us in fig. 88 centrifugal sifting machine constructed on one of the largest scales. In this cacao is introduced in the floor of the sieve through a feeder, and by means of an elevator. The sifting cylinder is spanned with silk or bronze gauze, and conceals in its interior a rough sort of preliminary sieve, the purpose of which is to prevent the larger unpowdered pieces penetrating to the silk gauze. There is a ventilator inside this rough sieve, which produces and transmits an air current, so that the meshes are kept open. Under proper guidance it is practically impossible for the machine to break down, although the sieve must be cleansed twice daily, an operation scarcely requiring more than two or three minutes, as it is not necessary first to remove the part under consideration. Because of this easy manner of cleaning, the centrifugal sifter far excels all others, as the plan sifter, the latter generally having to be dismounted before this operation can be proceeded with.

The powder issues from the first outlet of the sieve. There is a second, where both preliminary and cylinder sieve transmit their overflow, and this is then again conducted to the pulveriser in order to be worked up once more. Pulverisers and sifting apparatus can be so combined by means of conveyors and elevators that they work automatically, which is always of immense advantage where a large daily output is in question. But pressed cakes which

298

are to be conducted through the machine in broken pieces must first be treated in a preliminary crusher (cf. figs. 85a and b).

Fig. 88a shows one of the plansieves of the firm Baumeister, and protected by patent, which also finds employment for the sieving of cacao powder.

This machine possesses four round sieves lying one upon another, on which the material to be sieved is moved by a crank driving power just as on a hand sieve, so that the surface of the sieve is fully employed. The sieves possess neither projection nor hauling gear, the sieving is effected without pressure or friction, and the powder is therefore loose and woolly. A brushing arrangement revolves without any mechanism, driven solely by the peculiar movement of the plansieve, under the wholly flat sieves, and this brushing arrangement any cacao powder which may adhere to the sieve and so prevents a displacement of the tension, as far as possible.

Fig. 89.

In the following illustration we give as an example the arrangement of a pulverising plant with pulverisers (cf. fig. 86) for a second time.

The preliminary crusher receives the cakes, and then a conveyor brings the broken pieces along to the elevator, which in its turn feeds the filling box of the pulveriser, the connection between the two being established by a sliding platform. The discharged material succeeds on a landing where it is cooled down a little. A second conveyor brings it to the elevator of the sifting machine. Whilst the fine powder is taken up in barrels collectively introduced under the apparatus, the remainder of the cacao passes along to the conveyor first mentioned, is mixed with other broken pieces of cacao cake, and so returns to the pulveriser.

In reference to the Dutch method of disintegration, mention must be made of the process adopted by Moser & Co. in Stuttgart[129], where the cleansed, shelled and moistened beans are enclosed in a rotating drum, so that they can be subjected to the influence of ammonia and water vapour, produced from a solution of ammonium carbonate, which is passed through the hollow interior of the drum. The beans are then roasted and so freed of superfluous ammonia, after which follow in regular order the processes of grinding, defatting and pulverising.

After this description of the Dutch and other well-known methods of disintegration obtaining in the manufacture of cocoa powder, we shall now proceed to describe such of the remaining processes as seem to deserve mention.

c. Disintegration after Roasting.

The chief difference between the following methods of procedure and the Dutch and other processes previously referred to is that in the former the beans are neither impregnated with alkalis before nor during the roasting, but after it has been carried out, and the impregnation occurs sometimes prior, and at other times subsequent, to the expression of the fat. The several stages of treatment which proceed this process succeed each other in the same order as in the preparation of chocolate, cleansing, sorting, roasting, crushing, shelling and trituration following one after the other. But if the treatment with alkali is to take place before the fat is expressed, the cacao passes from the grinding mill direct to the apparatus in which it is subjected to the action of a solution of potash or some other alkali.

1. Disintegration prior to Pressing.

301

The system of impregnating the ground but as yet undefatted beans with alkali was first introduced into Germany by Otto Rüger, Lockwitzgrund. The principle features of the Rüger process are similar to those of other methods at present frequently met with, so that a detailed description would seem to be rather superfluous. Melangeurs may be conveniently employed in the treatment of cacao mass in a liquid state with alkalis, such as we have previously described, and illustrated in fig. 86 on page 210.

Fig. 90 a.

As preparing machines for disintegration, the kneading and mixing apparatus shown in working position in fig. 28, page 118, and in fig. 90 a with tilted trough for emptying are specially constructed and patented and quite deserve the popularity they have acquired.

Fig. 90 b.

Fig. 90 c.

Fig. 91 a.

Their construction and method of working are described on page 118. Other well-known machines for the purpose are the "Universal" mixing and kneading machines patented by Werner & Pfleiderer, which are shown in figs. 90 b and 90 c. As regards the general outlines of their construction, it will suffice to refer to the excellent descriptions of the machines which occur in the catalogues issued by this firm. Mention cannot fail to be made, however, of the circumstance that in these machines the evaporation of the alkaline solvent is also effected. The working of the kneading arms facilitates the escape of vapour from the mass and prevents overheating from contact with the walls of the apparatus. Underneath, the trough is provided with a double jacket, that is heated by steam.

To maintain connection of the steam and water pipes whilst the trough is reversed there are two flexible metal tubes. Both are screwed to the fixed pipes. For carrying away the vapour given off there is a tin plate cover to the trough, provided with a charging aperture and a channel inside to catch the moisture collecting on the cover and discharge it. When the machine is to be emptied, the cover is raised and a receiver adapted to the size and form of the machine is so placed that the charge can be diverted into it. The tilting of the machine is effected mechanically, and depends on the working of a lever. So as to prevent spurting of the liquid material when discharging, the stirring arms can be stopped for a time.

From this "Universal" Kneader and Mixer the special type "Vacuum Kneader", system Werner-Pfleiderer, is distinguished, as its name implies, by a vacuum arrangement. As seen on illustrations 91 a and b, this comprises a pyramid-like cover made of cast iron, and shutting down air-tight. which is provided with indiarubber caulking, and binding screws, and is fitted up for steam heating. It moves on the frame of the machine and is counterpoised with weights, so as to facilitate its raising and lowering. On the front part of the lid there is a small aperture paned in with glass, and opposite on the interior in a specially protected compartment occurs an electric light arrangement, which admits of the continual observation of the material during the working up processes. In addition, small quantities of cacao mass can be introduced on removal of the glass pane without lifting up the lid; so that the advantages of the aperture are twofold. The upper part of the cover tapers off into a suction pipe, which itself terminates in a flanged support intended as a finish to the conduit from the airpump.

Fig. 91 b.

The kneading trough of the machine is made of cast iron, provided with a false bottom, and fitted up for heating with hot water or steam to a pressure of 7 atmospheres, or for cooling down with cold water. By way of rapid discharging, the trough is counterpoised with weights, and can easily be tilted over by means of a hand winch. Its interior, as also the kneading shovels, are clean scoured, and the bearings of the shovels stopped with easily adjustable stuffing boxes. These stuffing boxes (German Patent) are so fitted in that no greasing substances whatever can penetrate to the cacao mass, which is of the highest importance, as in the case of the ordinary stuffing boxes grease is sucked up into the kneading trough by the action of the air pumps and the material contained in this so rendered impure. The steam and water conduit to and fro is effected by means of supple metallic hose, which follow the movement of the trough as it is tilted.

The vacuum kneading machines have acquired great importance in the manufacture of milk chocolates, where it is chiefly a question of reducing mixtures of cacao, sugar, and condensed milk to a requisite thickness. Lately the value of the machine has been regarded as consisting in the main of the possibility of preparing cacao under vacuum which it affords.

It is easy to understand that the treatment of the cacao under vacuum demands a much lower temperature and takes place in about half the time requisite for open machines, where it must be carried out against the constant and contrary influence of the atmosphere, apart from the fact that the vacuum kneader preserves the aroma far better.

The alkali solution used in disintegration may be prepared in vats fitted with draw-off cocks, or, in small factories, in glass carboys such as are used for the conveyance of acids. Of the fixed alkalis, potash is preferable, since it is a natural constituent of terrestrial plants and therefore of the cacao bean, and so its employment introduces no foreign ingredient. Magnesium carbonate seems to find favour in many quarters, but we consider it less suitable as being insoluble in water, and therefore can only be incorporated with the cacao mass in a state of suspension. It is sufficient to have a potash solution some 90 or 95 % strong, answering to the requirements of modern medical treatises.[130] The salt is soluble in an equal quantity of water.

In preparing the solution, the best plan is to dissolve a known quantity in from 3 to 4 times as much water at the temperature of the room and then by diluting with water reduce this composition to the required strength. As for each 100 kilos of cacao still undefatted from 2 to at the most 3 kilos of potash and from 15 to 20 kilos of water are required, this 2 or 3 kilos of the salt should be dissolved in about 10 litres of water and the solution after diluted with the remainder of the water.

In using volatile alkalis, which are nevertheless falling into disuse more and apparently no longer maintain their reputation, ordinary ammonium carbonate which may be easily obtained in powder form at any chemist's, or a solution of ammonia, such as spirits of sal-ammoniac, may be used. The former is easily soluble in about five parts of water. From ½ to 3 kilograms of ammonium carbonate are generally reckoned for every 100 kilos of undefatted cacao material, and this amount is dissolved in water, the whole of the salt being at once introduced into from 15 to 30 litres, as when smaller quantities are used there ensues a decomposition of the salt and one of the products of decomposition, the carbonate of ammonium, remains undissolved.

The spirits of sal-ammoniac operate much more effectively than the ammonium carbonate on account of their high percentage of ammonia, and so only a third as much of this substance may be employed, and generally even smaller quantities prove quite sufficient. Consequently 100 kilos of defatted cacao should be mixed with 0·5-1 kilo of ammonia solution (specific gravity 0·96), previously diluted with 20 or at the most 29 litres of water. The mixture should be prepared in glass carboys immediately before use, because of the volatility of ammonia.

In the treatment of the cacao, salt solution and cacao are together introduced into a melangeur, or better into the kneading and mixing machine, and the apparatus being set in working order, steam enters, and removes the quantities of water which have been added, as well as the volatile alkalis. Whether all the water has been driven off or no can only be judged from the consistency of the mass after treatment, and it is just this that renders the process of little value. The cacao material issuing from the machine must be just as liquid as when it comes out of the triturating mills, and so long as it appears as a glucose substance, which very often happens where unsuitable mixing machines are employed, so surely will it contain water, and this may lead to the growth of mould or to the cacao developing a grey colour when packed in boxes. If the cacao cannot be sufficiently dried in these machines, it must be transferred to some sort of drying plant (where the temperature is about 48 ° C.), and there deprived of its still remaining moisture.

When volatile alkali is used, kneading and mixing machines cannot very well be dispensed with, as they work up the cacao material much more thoroughly and admit of a better distribution of the ammonia than the melangeur or incorporator. In this case it is advisable that the entire process be carried out in some apartment separated from the other rooms of the factory, in order that the pungent smell of ammonia may not be communicated to other products, a further evil connected with this method of disintegration. At the same time provision must be made for the escape of the discharged gas through flues leading out into the open air.[131]

The treated cacao, when perfectly free from water and volatile alkali, then passes on to the press, pulveriser and sifting machine successively, the several operations being proceeded with exactly as described. In the original process of Rüger's, the defatted and disintegrated cacao is dried after it has been reduced to smaller pieces, and then mixed with fat in such proportions as seem requisite and desirable, so that it is possible in this method to re-imbue a disintegrated cacao with its original percentage of fatty contents.

2. Disintegration after Pressing.

In this process, which may no longer be adopted as far as we can ascertain the mechanically prepared beans are roasted, crushed and decorticated, then ground in mills, defatted, and finally the cakes are broken up into a rough powder and treated with alkali in the manner above described. Care must here be taken to use as little water as possible in dissolving the alkali. It is best to employ potash exclusively, for it has been found that the last traces of volatile alkali are extremely difficult to remove from defatted cacaos as decomposed by the solution, and there is no means of neutralising the ammonia without at the same time causing material damage to the flavour and aroma of the product treated.

The concentrated solution of alkali may be conveniently sprayed on the powder while the latter is subjected to a constant stirring, an operation best effected in the melangeur. The final drying is carried out in hot closets, provided with an effective ventilator suitable to the purpose. After it has been thoroughly dried, the cacao next succeeds to the pulverising and sifting processes.

Some methods of rendering cacao soluble remain to be mentioned, wherein no alkali whatever is used, and in which the disintegration is effected by means of either water or steam. The first process of the kind was invented by

312

Lobeck & Co of Dresden[132] in the year 1883. The cacao beans, either raw, roasted, decorticated, ground or otherwise mechanically treated are exposed to heat and the action of steam under high pressure in a closed vessel, then subsequently powdered and dried. The process has little to recommend it and has not been able to establish itself accordingly, for hereby the starch in the cacao is gelatinised, and acid fermentation is introduced, such as does not fail to damage the final product. Then again, there is a danger of the cacao becoming mouldy in the store rooms, after being treated by this process.

A second method, patented by Gädke, German Patent No. 93 394, 17 th. Jan. 1895, consists in disintegrating by means of water in a less practical manner. The roasted, decorticated but as yet unground beans are moistened with water, and subsequently dried at a temperature of 100 ° C. after which succeed the processes of grinding, defatting, pulverising and so forth. This process has also failed to establish itself to any effect.

In our opinion any one of these methods skilfully and properly carried out will yield a marketable, hygienic and wholesome product, though some of them can boast of their own particular advantages. This holds good for the so-called "Dutch" method in particular, though it is open to the objection that the cacao so prepared is combined with an extraneous product and that the combination remains right up to the moment of consumption. Considered from this point of view, disintegration with fixed alkalis is generally less advisable than the optional treatment with water or volatile alkali, but it may be taken for granted that each manufacturer had better decide the several details best adapted to his own particular outfit.

A well made soluble cocoa powder should have a pure brown colour, without any suspicion of grey, should be perfectly dry, and feel light and soft when finely divided, so betraying that property which the French designate under the term "impalpable The peculiar aroma of the cacao must be retained, and especially should the preparation be preserved from the slightest taint of any ammonia combination, its taste being kept pure and cacao-like, any hint of alkalinity indicating defect in the manner of disintegration. Over and above delicacy of aroma and taste, that characteristic described as "solubility" constitutes a main criterion of quality in the eyes of the consuming public. To ascertain that only an empirical test can be employed.[133] About 7·5 grammes of cocoa powder are introduced into some 150 grammes of hot milk or hot water contained in a graduated beaker, and then the quantity of sediment which sinks to the bottom of the vessel in a given time is noted. The more slowly a sediment is formed and the smaller it is, the greater the "solubility" of the cocoa.

If it becomes necessary to give the cacao an additional flavouring, the spices or ether-oils generally employed in the manufacture of chocolate may be used in the course of pulverisation, and shortly before sifting.

C. Packing and Storing of the finished Cacao Preparations.

Chocolate will keep in its original condition for years, when protected from atmospheric influence. It is therefore generally, and especially where the finer qualities are concerned, packed up immediately after it leaves the last process, and ornamented chocolates are previously varnished with an alcoholic solution of benzoin and shellac (see page 250).

The inferior qualities are usually packed in paper and wooden boxes, but the superior first in tin-foil and subsequently in paper. C o c o a p o w d e r arrives packed in parchment boxes as a rule, and also in cardboard or tin boxes.

Although packing in parchment or waxed paper is hygienically and economically more advantageous than tin-foil packing, the latter is nevertheless to be preferred, not only because it is a better preservative of the aroma evident in the spices added, but also because it prevents an evil which also in the end leaves its mark on cacao, when stored a very long time, to wit, the development of rancidity. This is explained by the fact that the tin-foil sticks to the chocolate, and so hinders the penetration of air.

According to an act dating from June 25th. 1897, and in force in Germany (Reichsgesetzblatt No. 22), metal-foil containing more than one percent of lead may not be used in the packing of snuff, chewing tobacco and cheese. What holds good for other articles of consumption must also apply to cacao preparations, when they are so packed that they come first of all into contact with metal-foil, and not with paper. Tin-plating also, containing in its coating more than 1 % of lead and in the soldering more than 10 % is also inadmissible in the chocolate industry. Although it is said that the whole of the tin-plate fittings made in Germany are constructed according to an imperial standard, yet it may occasionally so happen that cheap packing material does not correspond and answer to the legal requirements.

The manufacturer can only protect himself against possible prosecution for contravening or neglecting the articles of this act by obtaining a written guarantee as to the quality of the tin-plate supplied.

The rooms where chocolate wares are stored should not be too warm, and it is indispensable that they be kept dry, for heat accelerates the volatilisation of their aroma and also the rancidity to which cacao is liable, whilst moisture spoils the general appearance of the chocolate and promotes the growth of mould. This development of mould, which is first noticeable after long storage in damp, dark warehouses, is principally due to the growth of a fungus which Royer has named "Cacao-oïdium[134]

As the numerous wrappings (in tin-foil, paper, etc.) are at present only effected by hand labour, they mean an appreciable increase in the price of the goods. This is of less moment for the chocolate tablets as the small napolitains and the like. Therefore attempts have often been made to effect this wrapping by means of machines[135], and I have seen among others two models for napolitains, one on a large and the other on a small scale, the property of a Hamburg chocolate factory, and constructed by the firm of A. Savy & Co., Paris, which same machines were said to effect the wrapping in tin-foil, folding and additional packing in paper, as also the final closing, automatically and well; but just as I requested to be shown the machines, I was told that they were for the time being not in working order. Since then I have heard no more of the matter, and regret that the firm of Savy & Co., who have a branch in Dresden, have not been able to answer several letters which I sent them inquiring for further particulars. It must be that the machines have failed to answer their purpose, for otherwise they would have been assured of a hearty reception, no matter how dear they might have been. So for the nonce our chocolate packing must depend on hand labour.

Quite a different arrangement obtains in respect to cocoa powder, which was also originally packed up in paper bags by hand. This operation is to-day despatched in machines, as also in the case of other powder substances, like tooth-powder, dyes, patent foods, soap powder, etc., and this even in the smallest of factories. It is true that the machine built a decade ago by L. Wagner in Heilbronn and at that time described by Zipperer in our second edition, which was to wrap up a dozen packets simultaneously, seems to have failed, for it is no longer constructed; yet its place has been taken by a succession of other machines which have stood the tests of many years. The principle has been altered, many packets at one time not being filled, but always one only, and the advantage lies in the fact that the machine fills more exactly and with a higher degree of uniformity as regards the weights of the several packets.

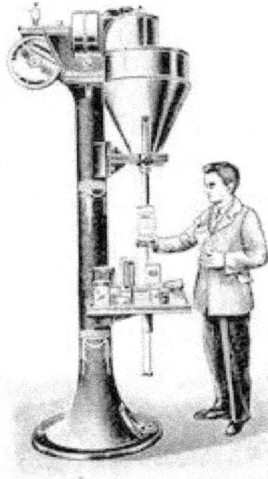

Figs. 92 and 93.

Apart from the "Machines for packing en masse" Co., Ltd. Berlin, who put out several automatic fillers, special mention may here be made of the firm of Fritz Kilian, whose automatic filler and packer "Ideal" (fig. 92) for quantities of from 25-2500 grammes, and "Triumph" (fig. 93), for quantities of from 1-100 grammes, have both long established their right to a place in every factory, their excellence being predominant.

Part III.
Ingredients used in the manufacture of chocolate.

A. Legal enactments. Condemned ingredients.

Chocolate is a mixture of cacao mass with sugar, to which usually spices and even cacao butter are also added. The sugar generally amounts to rather more than one half (60 percent) of the total mixture. Spices such as cinnamon, vanilla, cloves, nutmeg, mace, cardamoms, as well as cacao butter, or perfumes like peruvian balsam, are only added in small quantity so as to improve or alter the flavour as required. Recently, the ethereal oils of the spices have been used for this purpose as well as artificially prepared aromatic substances, such as vanillin, for example. Flour and starch[136], although the latter is seldom used, are permissible ingredients in cheaper kinds of chocolate but only when the fact of the addition is plainly stated. The kinds of flour usually employed are wheat and potato flours, rice-starch and arrowroot, dextrin and, less frequently, oat, barley, acorn, chestnut, or rye flour. In certain forms of dietetic chocolate, sugar being injurious to invalids, it is replaced by saccharin; another material, such as a leguminous flour from beans, peas or lentils, must be employed in its place.[137] In some kinds of fancy chocolate, harmless colours, tincture of benzoin etc. are used.

B. Ingredients allowed

I. Sweet Stuffs.

a) Sugar.

Both cane and beetroot sugar are employed in the manufacture of chocolate. As this naturally possesses a brownish colour, brownish white as well as white sugar is used for mixing with the cacao mass. The kinds of sugar used are:

1. Sugar dust, a white crystallisable and very fine powder.

2. Crystal or granulated sugar, consisting of loose, plain crystals, and suitable for almost all purposes in the manufacture.

3. Sugar flour I, II, and III which is a difficultly crystallisable sugar containing an amount of molasses increasing with the number, and it is of a more or less brown colour.

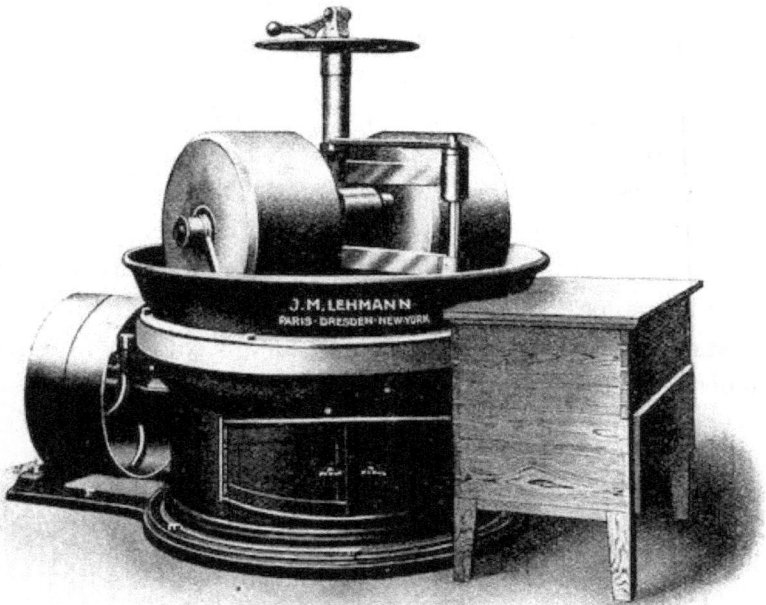

Fig. 94.

The chocolate manufacturer nevertheless requires the sugar to answer to certain characters. It must dissolve in half its weight of warm water forming a sweet syrup. The syrup must have no action on either red or brown litmus paper i. e. have neither acid nor alkaline reaction, and on no account coagulate boiling milk.

The sugar is usually added to the cacao mass in the form of a very fine powder and sometimes in a coarser condition, though that is not to be recommended. By using finely powdered sugar, the rolling of the cacao mass is considerably facilitated and the manufacture is accelerated. The sugar must be perfectly dry, as damp sugar yields a dull chocolate which readily crumbles.

Fig. 95.

For grinding the sugar, the so called edge-runner mill as shown in figure 94 was formerly employed.

It is like the melangeur constructed of a firmly fixed bed-stone and two cylindrical runners.

The pulverised material issuing from such an apparatus

must then be passed through one of the various kinds of sifting machines, where the finer parts fall through the meshes of a silken sieve, whilst the rougher are discharged at the end of the arrangement: for small factories such machines as the drum sifters illustrated in fig. 95, and for the larger those centrifugal sifters which have already been fully described.

The constructions for grinding have of late been considerably perfected. The most practical arrangements for pulverising all kinds of granulated sugar and so-called lump sugar, are those combined grinding and sifting installations such as are executed by the firm of J. M. Lehmann in Dresden. The grinding is here effected by disintegrators (revolving arms, etc.) similar to those used in the pulverising of cocoa powder as described on page 212. The output of these disintegrators[138] is extraordinarily large, and the harder and drier the ground sugar is, the finer the pulverised material resulting. We annex a diagram of the machine in fig. 96.

Fig. 96.

The granulated or lump sugar is filled into the hopper
and thence lead along a conveyor to be ground in another
part of the machine, and can be controlled as regards
quantity. The blades, which pass through about 3000
revolutions a minute, seize the sugar and swing it against
the ribbed walls of the mantle, after which it falls in smaller
fragments on a grater fitted in the under part of the
apparatus. The sugar which passes through the grating is
now conducted by conveyor and elevator to the sifting
arrangement, whilst the rougher material is again whirled
round by the blades. This sifting arrangement consists of a
cylindrical sieve, on the interior of which there occur
revolving arms which provide for the despatch of material

through the various sieves. The rougher stuff which remains is removed by hand or some other mechanical means and transported to the hopper once more. A chamber placed above the machine and connected with the grinding apparatus by means of pipes provides for the protection of the machine against dust.

Such installations are constructed in various sizes and fashions, and possess immense outputs (up to even 5000 kilogrammes daily). That they must be built in special shops is clear from the fact that so large a quantity of dusty sugar sacks need transporting after the processes are completed. It is further to be noted that the fineness of the sugar corresponds to the mesh-work of the sieves, which as we have previously stated, can be chosen with any size of hole desired, yet this naturally influences the machine, and recently a very high standard of fineness has been generally dropped, and rougher siftings are now made, as when the sugar is too fine. — e. g. in the case of the cheaper qualities — it absorbs too much of the fatty contents, and so necessitates the addition of cacao butter, whilst on the other hand, when the chocolate is of a finer quality, the sugar is sufficiently reduced in the trituration to which the mixed material is subjected.

b) Saccharin and other sweetening agents.

Apart from the sugar, which is such an important factor in the chocolate manufacture, mention must also be made of another sweetening material, formerly frequently used as a substitute for sugar, but now only to be obtained at the apothecary's on exhibition of a medical order, in consequence of certain legal restrictions which have recently come in force. It is called Fahlberg saccharin, and again zuckerin, sykorin, crystallose, "Süßstoss Höchst" and sykose.

Saccharin is not like sugar a carbohydrate naturally produced by plants, but a derivative of the aromatic compounds which the chemist has artificially constructed from the products of the distillation of coal.

Saccharin is benzoyl-sulphonimide, and it has the chemical formula

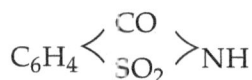

$$C_6H_4 \underset{SO_2}{\overset{CO}{<}} > NH$$

It is a white, crystalline powder, so exceedingly sweet that its taste can be perceived in a dilution of 1 in 70000. It is only slightly soluble in cold water (1: 400) but more easily so in hot water (1: 28). The material known as easily soluble saccharin is its sodium salt. It contains 90 percent of saccharin and is the most easily digested compound of saccharin.

For technical, domestic and medicinal purposes the soluble saccharin which is only from 300-450 times as sweet as sugar is employed. Besides being unfermentable saccharin has very slight antiseptic properties; according to L. Nencki[139] the digestibility of albumin is less affected by it, in the proportion usually added to articles of food, than by Rhine wine, or by a sugar solution of equal sweetness. Saccharin is entirely unaltered in the human organism, hence it forms a welcome sweetening material for invalids suffering from diabetes, corpulence or diseases of the stomach to whom ordinary sugar is injurious. The substances known as dulcin and glucin are analogous to saccharin in sweetening property, the first being phenetol-carbamid and the latter a monosulphonate of amido-triazine.

The latest substance of this class is termed "sucramin" and consists of the ammonium salt of saccharin. It is readily soluble in water, less so in alcohol and is 700 times sweeter than sugar. It can be obtained either in the pure form or mixed (20 percent) with sugar.

In chocolate making, saccharin is at present of little importance, owing to the relatively small volume required as compared with sugar. Recently it has again been recommended to the extent of 0·76 percent as a sweetening material for cocoa powder. It would certainly be of value in cocoa powders to be consumed by invalids and persons not able to take sugar, although it will never come into general use. The detection of saccharin has acquired increased importance in Germany since the passing of the acts of October 1st 1898 and July 7th 1902, regulating the trade in artificial sweetening materials. According to Zipperer's experiments, it may be detected in the following manner: A mixture of 5 grammes of the finely powdered substance with 100 ccm of water is allowed to stand for 2 hours, occasionally stirred and afterwards filtered. The filtrate is acidulated with three drops of hydrochloric acid and evaporated to 20 ccm, then shaken[140] with 50 ccm of ether in a separator and left standing for a day to separate into two layers. The ether solution is separated and evaporated to dryness in a beaker, the residue being mixed with 0·1 gramme of resorcin and 4-5 drops of concentrated sulphuric acid[141] (Börnsteins test). The mixture is then heated over a small Bunsen flame and the melted material saturated with normal sodium hydrate. The appearance of a strong fluorescence indicates the presence of saccharin. Saccharin can also be easily recognised by the sweet taste of the ether residue.

II. Kinds of Starch, Flour.

The chief kinds of starch used in chocolate making are rice starch, arrowroot, potato starch and wheat starch, occasionally also small quantities of dextrine.

1. Potato starch or flour.

Potato starch is a white or faintly yellowish powder in which single, glistening granules can be seen by the naked eye. Under the microscope the granules appear mostly single with evident striae, usually with pointed ends containing the nucleus; they are also eccentric in structure. This starch rarely contains fragments of tissue. It is prepared by first treating finely divided pared potatoes with 1 percent dilute sulphuric acid, then washing, drying and grinding the starch.

2. Wheat starch.

Wheat starch can be obtained either from crushed wheat or from wheaten flour by treatment with water after the nitrogenous constituent, gluten, has been separated by kneading. It amounts to about 60-70 percent of the grain. Under the microscope the granules appear to differ considerably in size. They are distinguished from potato starch by the nearly central hilum, surrounded by faintly marked concentric striae, and again by the granules being more frequently adherent. Wheat flour rather than the starch is generally used in chocolate making.

3. Dextrin.

When starch is heated to between 200° and 210° C. it is converted chiefly into dextrin or starch gum with a little sugar. Dextrin is a white to yellowish and tasteless powder with a peculiar smell; it differs from starch in being readily soluble in water. It gives a reddish colour with an aqueous solution of iodine. Fehling's solution is unaffected by dextrin in the cold, but on long continued heating it is reduced to red cuprous oxide.

4. Rice starch.

Rice starch is obtained from inferior kinds of rice and from rice waste by treatment with water. It appears under the microscope as small granules or oval bodies of various sizes. According to their position the granules always seem to be polygons,[142] formed by coalescence. It is thus easily distinguished from the previously mentioned starches.

5. Arrowroot.

Several kinds of starch, obtained from the tubers of various species of plants are commercially known under this name.

1. West Indian arrowroot, from Maranta arundinacea, is a fine and almost white powder. Under the microscope it always appears to consist of pear or spindle-shaped granules with eccentric hilum.

2. East Indian arrowroot is obtained from various species of ginger plants. It is a fine white powder and is seen under the microscope as single granules with well marked eccentric hilum and closely stratified at the spindle-shaped ends. It much resembles Guiana arrowroot, which is obtained from varieties of Yam.

3. Queensland arrowroots from species of Cycas and

Canna, appear as flat, coarse and mostly single granules. They can be easily distinguished from other kinds of starch by the large size of the granules.

4. Brazil arrowroot, from the Manihot plants which belong to the order of Euphorbiaceae. Under the microscope the granules appear compound, the parts being of a drum or sugar loaf shape with many concentric striae.

6. Chestnut meal.

Chestnut or maron meal also comes under consideration in the chocolate industry. The appearance of the starch granules is most characteristic. They are partly single and partly composed of two individual granules. The single granules, according to J. F. Hanausek[143], appear in such a variety of forms as to defy a summarised description. Frequently they occur oval, spindle, club, or flat kidney shaped, resembling those of the leguminous family; but especially to be noticed is the triangular contour of some granules, as well as some with projecting points. The central nucleus and its cavity are generally distinct, but the stratification is very slight or quite unrecognisable.

7. Bean meal.

Of the leguminous meals that of beans is chiefly used as an adjunct in cocoa powders and chocolate, sweetened with saccharin, on account of its relatively large proportion of albuminous substance and small amount of starch. The meal is generally obtained from the seed of the common white bean. (Phaseolus vulgaris.) The starch granules under the microscope appear oval or long kidney shaped, with distinct nucleus cavities and furrows, as well as a distinctly marked stratification. Their length averages from 0·033 to 0·05 mm. The meal has a disagreeable leguminous taste when cooked, but that disappears when the meal is slightly roasted.

8. S a l e p.

Salep which is now very seldom used as an admixture to chocolate (Rakahout of the Arabs)[144] is an amylaceous powder prepared from the tubers of various kinds of orchids. Under the microscope salep appears as fairly large translucent masses which consist of an agglomeration of very delicate walled cells giving the starch reaction with iodine.

III. Spices.

a) General Introduction.

We cannot too strongly recommend the manufacturer to pulverise the spices, e. g. cinnamon, cloves and the like, himself, for such as are bought ready pulverised have frequently been adulterated with admixtures of wood, flour or bark. This is the more essential as sometimes pulverised cinnamon is distilled with steam to obtain an extract of its ethyl oil, and then the residue, which is of considerably inferior value as regards aroma, sold as genuine cinnamon powder. Such adulteration can neither be demonstrated under the microscope nor chemically, so that it is impossible to protect oneself against them.

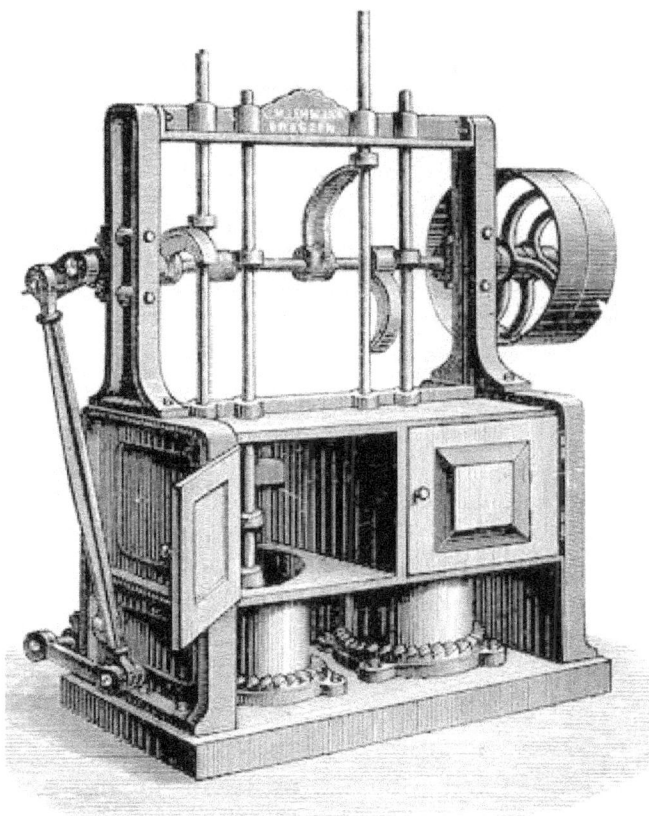

Fig. 97.

The edge runner mill and sieving apparatus described in
connection with the pulverising of sugar also adapt
themselves to reducing spices, although generally other
machines are used for this purpose, either the well-known
ball mills[145] consisting of a hollow spherical ball revolving
round its axle, inside which the spices are shaken, crushed
and completely pulverised by the action of a number of
heavy metal balls, or in other cases pulverising mills and
stamping arrangements proper.

Fig. 98.

The following stamp arrangement, shown in fig. 97, is very practical in the pulverisation of all manner of spices, and is driven by a force of 1·5 H.P. The strong frame, which is walled in with iron, is dust-proof. Whilst the stamper is being raised, the pots are revolved round their axles, and so the substances to be pulverised are mixed together. Other machines much used in pulverising are seen in fig. 94. Another smaller pulverising mill is pictured in fig. 98. This machine is adapted for a middle sized production. The grinding arrangement in which the pulverising takes place is conically built and is made completely of granite; the regulation is effected by means of a working beam, the batting arm of which is fitted on to the upper part of the apparatus. A sieving of the material to be pulverised does not generally take place in this machine. For small production for example for confectioners who manufacture chocolate also incidentally, one can also use the machines pictured in the figs. 95 & 99, the method of working of which may be at once understood. The different degrees of fineness of the material to be pulverised are reached by passing the powder through drum sieves of different widths of mesh and all the sieves are set in motion at the same time by the machines.

Vanilla.

Only the most important features of the spice so valuable in chocolate making will be noticed, since the characteristic aroma of the true vanilla has been to a large extent supplanted in practice by artificially prepared vanillin.

Vanilla is the fruit capsule of an orchid, Vanilla planifolia, which is generally cultivated with the cacao tree, as the same climate and soil suit them equally. According to Möller, the shoots of the vanilla are fastened to the cacao tree, on the bark of which they soon strike root. The aerial roots and tendrils then put forth fleshy leaves, in the axils of which arise large odourless and dull coloured flowers which yield after a lapse of two years long thin capsules. The capsules are filled with a transparent balsam, in which the black seeds are imbedded. It is in the balsam that the vanillin, which gives vanilla its unequalled aroma, is produced. The fresh gathered vanilla fruit (see the investigations of W. Busse[146] contains no free vanillin or merely an infinitesimal quantity.

Fig. 99.

It is rather developed by subsequent treatment in which heat appears to be necessary. Vanillin, like cocoa-red and theobromine, is formed by the splitting up of a glucoside by fermentative action. In some kinds of vanilla, piperonal, an aromatic body, which occurs in larger quantities in Heliotropium europaeum and peruvianum, has also been observed.

The commercial kinds of vanilla come from Mexico, Tahiti, Réunion, Mauritius, Mayotte, Seychelles, Ceylon and Java, which in 1891 produced respectively:

Réunion (Bourbon)	50-65,000 kilos
Mexico	55,000 "
Mauritius	13-15,000 "
Mayotte (Comoro Islands)	8-10,000 "
Seychelles	4- 6,000 "

334

The best commercial kinds of vanilla come from Mexico, Bourbon, and Mauritius, and command a higher price than the other kinds. The quantity is gauged by the length (10-24 cm), and plumpness of the pods. Fine quality is fatty and dark coloured, inferior quality is dry and reddish. The outside of the pods in the Bourbon vanilla, contains highly esteemed vanillin crystals, which are wanting in the Mexican variety. Vanilla flowers in October and November, is gathered in the following months of May, June, and July, and is prepared in October and November. At the beginning of November the first instalment of the new harvest arrives in Marseilles, which is the chief commercial place for vanilla. The most important operation, in preparing vanilla is to attain the proper degree of dryness. This is arrived at nowadays by the use of calcium chloride. The pods are first placed in a metallic box lined with wool which is placed in warm water so as to superficially dry them; they are then transferred to a suitable constructed drying closet containing calcium chloride and allowed to remain there for 20-30 days. 100 pounds of vanilla are reckoned to require 40 pounds of calcium chloride. The great advantage of this process is that the fruit, so dried, better retains its aroma.[147] Insufficiently dried vanilla does not keep, but soon becomes mouldy, whilst overheated vanilla keeps well, but is brittle, breaks easily and consequently has little commercial value. Vanilla covered with mould (A s p e r g i l l u s   r e p e n s and M u c o r   c i r c i n e l l o i d e s) is sought to be improved in various ways and is sold as of inferior quality.[148] It is worth observing that those persons who in the course of business handle vanilla show characteristic symptoms of poisoning. It affects the eyes and nervous system and produces eruptions on the skin. The complaint, however, is not of a dangerous nature, for the workmen quickly become accustomed to vanilla so that, after recovering from the first attack, they can resume work without risk to health.[149]

On account of its high price, vanilla is much subjected to adulteration; either by an admixture of the more cumarin-smelling vanillin (Pompona or La Guayra Vanilla [Vanilla Pompona Schieder]) or other less valuable vanilla fruit; sometimes pods that have been deprived of vanillin by extraction with alcohol are used for that purpose; their colour and appearance being restored by immersion in tincture of benzoin and coating with crystals of benzoic acid, powdered glass etc. In doubtful cases of adulteration the vanillin must be quantitatively determined.

That can be done by W. Busse's method[150], in which the vanilla is extracted with ether in a Soxhlet's apparatus. The extract is shaken with a solution of sodium bisulphite, the vanillin then set free with sulphuric acid and the disengaged sulphurous acid removed by a stream of carbon dioxide. The vanillin is then shaken out with ether and on evaporating off the ether, vanillin is left in a pure condition. Busse found by this method in East African vanilla 2·10 percent of vanillin, in the Ceylon 1·48 percent, and in the Tahiti variety from 1·55 to 2·02 percent. In America the so-called vanilla extract, instead of vanilla, is used and it lends itself to adulteration much more easily than natural vanilla. William Hesse has given methods and results obtained in the investigation of the extract.[151]

5. Vanillin.

Vanilla in the chocolate industry has recently been almost entirely superseded by the use of artificially prepared vanillin, which serves as a complete substitute for the essential and valuable constituent of vanilla. In comparing vanillin with vanilla, regard must be had to the amount of vanillin in the latter, which may vary to the extent of 50 percent according to whether the vanilla was damp, dry, fresh or stored. The finest kinds of vanilla seldom contain more than 2 percent of vanillin and in many kinds it varies between 0·5 and 2·5 percent. It may also happen that vanilla with 0·5 to 1·0 percent may be equally as fine in appearance as one of high percentage, hence the aroma value must be taken into consideration. In addition to possessing a uniform and permanent perfume vanillin is cheaper in price.

Vanillin occurs naturally not only in vanilla but also in very small amount in certain kinds of raw sugar, in potato skins and in Siam benzoin; it can be produced artificially from coniferin which is obtained from pine wood, or by the oxidation of eugenol, a substance contained in oil of cloves, from both of which Tiemann and W. Haarmann[152] first prepared it in 1872. In the course of the last ten years a number of processes have been discovered whereby vanillin can be artificially produced. The reader who is interested in this subject will find it fully discussed in a paper by J. Altschul in No. 51 of the Pharmazeutische Centralhalle 1895.

The competition which arose through the processes of Haarmann and Reimer of Holzminden and G. de Laire of Paris, whose products owing to patent rights had controlled the market from the commencement, produced a steady decrease in the price of vanillin.

The following table drawn up by J. Rouché[153] shows the revolution in price which has occurred in this article and how, in the course of time, a small business with large profits has been transformed into a large business with small profits.

The variation in the price of vanillin
Marks per Kilo.

1876	1877	1878	1879	1881	1882	1884	1885

7000	4000	2400	1600	1600	1600	900	900
1886	1888	1890	1892	1893	1895	1897	
700	700	700	700	700	560	108	

The chemical formula of vanillin is $C_6H_3(OCH_3)$ (OH)CHO; it melts between 82-83 ° C. and sublimes at 120 ° C. The colourless four-sided crystals have a strong vanilla odour and taste, are difficultly soluble in cold water, easily in hot water and very readily soluble in alcohol.

Vanillin is much adulterated. Cumarin, the aromatic principle of the melitot (m e l i o t u s o ffi c i n a l i s) and of tonquin beans etc., can be prepared cheaply and it is fraudulently used in large or small quantity to imitate the vanillin aroma. A sample of vanillin bought in Switzerland was found by Hefelmann[154] to contain 26 percent of antifebrin. The American "vanilla crystals" consist of a mixture of vanillin and antifebrin, or vanillin, cumarin and benzoic acid; latterly that article is stated to consist only of cumarin, antifebrin and sugar.

The melting point of genuine vanillin is a characteristic indication. Admixtures of vanillic acid and antifebrin cause depression of the melting point (4-8 ° C. according to the amount and character of the two substances [Welmans])[155]. For the quantitative determination of vanillin in mixtures, Welmans takes advantage of its behaviour towards caustic alkalis, with which, like phenol, it forms compounds that are easily soluble in water, but sparingly so in alcohol. The process is as follows: 1 gramme of the substance is placed in a cylinder of 200 ccm capacity with 25 ccm of alcohol, 25 ccm of approximately semi-normal alcoholic potash and 2 or 3 drops of phenolphthalein solution and agitated until completely dissolved. The excess of alkali is then titrated with semi-normal hydrochloric acid, and, at the same time, the strength of the alcoholic potash after adding 25 ccm of alcohol is ascertained. The number of cubic centimetres consumed is multiplied by 0·076, the semifactor for vanillin. In the case of vanilla sugar, 10 grammes are treated with 50 ccm of water to dissolve the sugar, then the alcoholic potash is added and the operation carried out as before described.

1 gramme of vanillin requires 6: 58 ccm of normal potash (= 0·36842 g KOH).

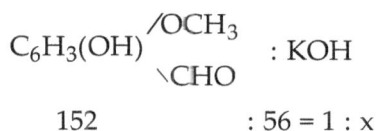

$$C_6H_3(OH) \begin{matrix} \diagup OCH_3 \\ \diagdown CHO \end{matrix} : KOH$$

$$152 \qquad\qquad : 56 = 1 : x$$

If cumarin is suspected to have been added to the vanillin it can be detected and separated, according to Zipperer's experiments, by the method of W. H. Hess and A. B. Prescott.[156]

The substance is dissolved in ether and the solution shaken up with a weak solution of ammonia. The vanillin will be found in the aqueous layer in the form of an ammonium compound, whilst the cumarin will be dissolved by the ether. The vanillin can be identified by the sandalwood oil reaction as described by Bonnema,[157] and the cumarin can be determined by direct weighing.

The financial advantage in using vanillin in place of vanilla is apparent. The average price of vanilla is now 45 to 50 shillings per kilo. But as 25 grammes of vanillin are equal in perfume to 1 kilo of vanilla and, at the rate of 35 shillings per kilo, that quantity costs only 10½ vanilla is nearly sixty times dearer than vanillin. The consumption of vanillin has increased to an enormous extent, and in the United States Henning has estimated the consumption during 1897-1898 at over 100000 ounces. The same author points out the remarkable fact that this enormous consumption of vanillin has scarcely any effect on the demand for vanilla pods, the market value of which is not only maintained but has a tendency to increase.

In order to have it in a finely divided condition, as required for the factory, it is recommended to rub the vanillin down with sugar, in the proportion of 100 grammes of vanillin to 2 kilos of sugar, in the following manner; 100 grammes of vanillin are dissolved in 500 grammes of hot alcohol and this solution added to 2 kilos of finely powdered sugar; then the whole is placed in a rotatory comfit boiler and dried by a blast of warm air at 40 ° C. Whilst vanilla must be very carefully packed that it may not become mildewed and deteriorate, vanillin on the other hand keeps very well in such mixtures so long as they are kept from damp, which might cause the sugar to ferment and thus gradually decompose the vanillin.

d) Cinnamon.

There are three commercial kinds of cinnamon in Europe.

1. Ceylon cinnamon, which represents the finest kind, is the bark of C i n n a m o m u m   c e y l a n i c u m, a native of the island of Ceylon. The bark is very light and brittle, seldom more than 0·5 mm thick, externally yellowish brown with long stripes, whilst it is somewhat darker on the inside. Its fracture is short and fibrous, and a traverse section shows externally a sharply defined light colour with a darker inside zone.

2. Cassia or Chinese cinnamon is from C i n n a m o m u m C a s s i a, a tree which grows wild in the forests of Southern China. The bark is thicker than that previously described, often 2 mm thick. It is in single tubes, harder and thicker

340

than the Ceylon kind, with frequently adherent fragmentary tissues of the corky layer. The colour is a greyish brown, the fracture even, with a light zone in the section.

3. Malabar or wood cinnamon consists of the less valuable kinds and is derived from different varieties of cinnamon trees which have been planted in the Sunda and Phillipine islands. In appearance it resembles the Chinese more than the Ceylon cinnamon.

The aromatic taste of cinnamon is due to the ethereal cinnamon oil which, in Ceylon cinnamon, amounts to 1 percent; the ash should not exceed 4·5 percent. An ethereal oil is also present (about 1·8 percent) in the leaves of the Ceylon cinnamon tree, but it is quite different from the bark oil, resembling in its properties more the oils of cloves and pimento. On account of its penetrating odour and pungent taste its employment in chocolate making is little to be recommended.

It cannot be too much insisted on that with spices like cinnamon, cloves, etc. the manufacturer should grind them himself and not purchase them in fine powder, as the latter is frequently adulterated with admixtures of wood, meal bark, etc. This is more to be recommended as ground cinnamon has frequently been deprived of the ethereal oil by distillation with steam and the bark then flavoured with a small amount of cinnamon oil and sold as powdered cinnamon. Such an adulteration can be detected neither chemically nor microscopically.

e) Cloves.

Cloves are the incompletely developed flowers of the clove tree, C a r y o p h y l l u s a r o m a t i c u s of the Myrtaceae.

The most important commercial kinds are the Zanzibar, Amboyna, and Penang cloves. The aromatic principle of cloves is an ethereal oil which they contain to the extent of 18 percent. The adulteration of cloves is much the same as in the case of cinnamon. Genuine cloves should not give more than 6 percent of ash.

f) Nutmeg and Mace.

Nutmeg is the seed kernel of the fruit of M y r i s t i c a m o s c h a t a known as the nutmeg tree, which is indigenous to Malacca. In the thick pericarp of the fruit, resembling the apricot, is found the brown seed surrounded by a deep red reticular mantle. This last is the seed mantle or arillus and when separated from the kernel is known commercially as mace.

The furrows on the surface of the nutty seeds are filled with a white mass which consists of lime, in which the nuts have been laid after drying in order to protect them from the attack of insects. The aromatic constituent of nutmeg and of mace is also an ethereal oil. The seeds contain 8-15 percent of ethereal oil with 25 percent of a fatty oil; mace contains 4-15 percent of ethereal oil and 18 percent of fatty oil. As both spices occur in commerce in whole pieces, adulteration is not to be feared.

g) Cardamoms.

Of these there are two kinds on the market:

1. The small or Malabar cardamoms.
2. The long or Ceylon cardamoms.

Both are the fruit, although very different in form, of a

species of the ginger plants which is indigenous to Ceylon and Malabar.

The Malabar cardamom is three cornered oblong and about 1 cm in size. In the fine brown pericarps are enclosed, adhering together, 6-8 angular seeds, 3 mm in size, having a pungent aromatic taste.

The Ceylon cardamom is four times larger than the Malabar kind. The grey brown pericarp encloses about 20 dark greyish brown seeds about 6 mm large. The aroma of the Ceylon cardamom is due to an ethereal oil which it contains in quantities sometimes reaching 6 percent. Madras and Malabar cardamoms contain 4-8 percent of ethereal oil. As the Ceylon cardamoms are cheaper than the Malabar kind a confusion of the two seeds might possibly be to the disadvantage of the buyer, but the above description of their relative size would suffice to distinguish them.

Exact accounts of the characteristic properties, the chemical and microscopical investigation as well as of the impurities and adulterations of the materials previously mentioned as being used in cacao preparations are to be found in volume II of the "Vereinbarungen zur einheitlichen Untersuchung und Beurteilung von Nahrungs-und Genußmittel sowie Gebrauchsgegenständen für das Deutsche Reich"[158] to which those who desire further to investigate this subject are referred.

IV. Other Ingredients.

a) Ether oils.

As previously remarked in the case of vanillin, it is becoming more and more the custom to substitute perfume substances for powdered spices. This practice is quite justified since the entire perfume of a spice is made use of and the worthless woody and indigestible fibre is thus excluded from the finished preparation.

The following are the ether oils used in practice:

1. Cinnamon oil,
2. Clove oil,
3. Cardamom oil,
4. Coriander oil,
5. Nutmeg oil (ethereal),
6. Mace oil (ethereal).

The amount of ether oil that should be used in place of the corresponding spice is a matter of taste. The maximum percentage of the oil in the respective spice might serve as a standard, as for example in the case of cinnamon oil, which is contained in the bark to the extent of 1 percent, about the hundredth part of the oil would be required to correspond with the prescribed weight of the bark. But as the yield of oil from one and the same kind of spice varies to a considerable extent according to season and locality, the percentage value can only be used as a general guide, and the final decision must be always regulated by the taste.

The ethereal oils can be incorporated in the cacao preparations (mass, powder etc.) either in a spirit solution or ground down with sugar. The latter method is naturally only used when sugar is to be added to cacao preparations. To prepare the alcoholic solution 10 parts of the ethereal oil are dissolved in 90 parts of strong alcohol. The mixture of oil with sugar can be made by triturating 2·5 parts of the ethereal oil with 100 parts of sugar in a porcelain mortar and grinding down with the pestle until the sugar and oil are intimately mixed. Of the alcoholic solution it is necessary to take 10 parts, and of the oil-sugar 40 parts to one part of ethereal oil.

II. Peru balsam and Gum benzoin.

Peru balsam is at present very much used as a perfume in chocolate making. It is obtained from the Papilionaceous Myroxylon Pereira which is indigenous to the western part of Central America. It is a thick, brownish black, liquid balsam which in thin layers appears transparent and has a peculiar smell and burning taste; it is almost completely soluble in alcohol, chloroform, and acetic ether. The aromatic substance of this balsam is cinnameïn, which consists essentially of the esters of benzoic and cinnamic acids and benzyl alcohol together with an alcoholic body "Peruviol", which has the smell of honey. In addition to cinnameïn (71-77 per cent) the balsam also contains a resin ester (13-17 percent). According to K. Dieterich, Peru balsam is the better for containing more cinnameïn and less resin ester. Peru balsam is adulterated with fatty oils, copaiva, gurjun-balsam, storax, colophony, turpentine, and tolu balsam. In regard to the chemical investigation of this balsam the work of K. Dieterich[159] may be consulted.

The Sumatra benzoin is the most important of the commercial kinds for chocolate making. It is obtained from one of the Styracae, Styrax benzoin, and is a reddish grey mass in which separate tiers of resin are embedded. Benzoic acid and vanillin are the most important constituents. It is adulterated with Palembang benzoin, colophony, dammer, storax, and turpentine. Respecting the chemical investigation of commercial benzoin the above-mentioned work of K. Dieterich may also be referred to.

Benzoin is almost exclusively used for the preparation of

chocolate varnish and sweets laquer, which are prepared by dissolving from 25 to 45 grammes of the laquer body in 100 grammes of strong spirit. The laquer body may contain varying quantities of benzoin and bleached shellac. The decorations of chocolate are painted with this laquer in order to give them a glistening appearance and greater durability.

V. Colouring materials.

The following colouring materials are permitted by the German law of the 14th May 1879 to be used for sugar goods and consequently also for chocolate and cacao preparations.

W h i t e: finest flour starch.
Ye l l o w: saffron, safflower, turmeric.
B l u e: litmus, indigo solution.
G r e e n: spinach juice as well as mixtures of the permitted blue and yellow colours.>
R e d: carmine, cochineal, madder red.
V i o l e t: mixtures of the harmless blue and red colours.
B r o w n: burnt sugar, licorice juice.
B l a c k: chinese ink.

In the meantime a number of comparatively harmless aniline colours have been permitted in Austria for colouring sugar goods and liqueurs, and eventually also for cacao preparations.[160] As in the author's opinion there is no ground for objecting to their use in other countries, a list of them is given under their commercial and scientific designations.

Red: Fuchsin = Rosaniline hydrochloride, soluble in water and alcohol.

Acid Fuchsin or Fuchsin S or Rubin = Sodium or calcium acid salt of rosaniline disulphonic acid, soluble in water.

Rocellin or Roscellin (Fast Red) = Sulpho oxyazonaphtalin, soluble in water.

Bordeaux and Ponceau red = product of the combination of β naphtol-disulphonic acid with diazo-compounds of Xylol and higher homologues of benzol, soluble in water.

Eosin = Tetrabrom-fluoresceïn, soluble in water and alcohol.

Phloxin = Tetrabromo-dichlor-fluoresceïn, soluble in water.

Erythrosin = Tetra iodio-fluoresceïn, soluble in water.

Blue: Alizarin blue = Dioxyanthraquinone-quinoline, slightly soluble in alcohol.

Aniline blue = Triplienylrosaniline, soluble in alcohol.

Water blue = Triphenylrosaniline, sulphonic acid soluble in water.

Induline = Azodiphenyl blue sulphonic acid and its derivatives, soluble in alcohol.

Yellow: Acid yellow R or fast yellow R = Sodium amidoazobenzol-sulphonate, soluble in water.

Tropaedlin 000 or Orange I = Sulphoazobenzoll α-naphthol, soluble in water.

Naphtholyellow = Sodium salt of dinitro-α-naphthol sulphonic acid, soluble in water.

Violet: Methylviolet = Hexa-and penta-methylpararosaniline hydrochloride, soluble in water and alcohol.

Green: M a l a c h i t e   g r e e n   =   Tetramethyl-diamidotriphenyl-carbinol hydrochloride, soluble in water and alcohol.

The above, as well as the following colours: (blue) amaranth, brilliant blue and indigosulfone, (red) erythrosin, also acid yellow S, orange L and light green S F, have in the meantime been accepted by the American Foods Act as perfectly harmless for colouring any and all articles of food.[161]

For some time past E. Merck of Darmstadt has supplied a perfectly harmless green colouring material under the name of chlorophyll, in alcoholic and in water solutions, as well as technical chlorophyll, for colouring oils and fats, which is the unaltered leaf green and is the best green colouring agent for articles of food and therefore for cacao preparations.

The chlorophyll which is soluble in fat has also been recommended like some of the aniline colours which are soluble in fat, as for example: Indulin 6 B (blue), Sudan yellow G, Sudan III (red), and Gallocyanin (violet) for colouring cacao butter; but in regard at least to the aniline colours mentioned, no authoritative sanction for their use has yet been given.

Part IV.
Examination and Analysis of Cacao Preparations.

A. Chemical and microscopical examination of cacao and cacao preparations.

The following observations will serve as an introduction to the chemical and microscopical examination of cacao preparations calculated to be of special value to the food chemist, corresponding as they do to the state of scientific progress at the present day and special attention being paid to the critical treatment of the methods of analysis etc. adopted.

a) Testing.

This is a point of great importance, inasmuch as it directly influences the result of the analysis of cacao goods. This is especially the case when dealing with c o c o a p o w d e r s as the test is liable to vary considerably according to the amount of moisture contained in the preparation and the degree of fineness of the powder. In the case of cocoa powders, the sample should be taken repeatedly from a large supply, and from all parts of the material to ensure getting an average sample. The samples taken should be of uniform volume and should, before proceeding to apply the test, be closely mixed together, being, if possible, first passed through a fine sieve. The material ready for the following experiments should then be placed in tin, or better still, glass receptacles with well-fitting corks or stoppers. Paper wrappings or cardboard-boxes are not to be recommended, as the powder is apt to become drier or moister according to the state of the atmosphere to which the packets are exposed.

The most suitable quantity for experimental purposes is, in the case of both chocolate and cocoa powder, as well as butter and covering material, 100 kilogrammes. When determining the amount of foreign fat in cacao preparations, however, as well as estimating the ash content of powder, up to 250 kilogrammes of sample material can be used. In Germany the regulations of the Commercial Agencies of the government public food chemists obtain when sampling and analysing cacao preparations.[162]

b) Chemical Analyses.

The analyses of all cacac preparations from a chemical point of view are conducted, almost without exception, with the object of determining the values for m o i s t u r e —m i n e r a l m a tt e r (estimation of the amount of the carbonic acid alkalis and the silicic acid)—f a t (estimation of foreign fat)—t h e o b r o m i n e and c o ff e i n e—s u g a r —s t a r c h (foreign starches)—a l b u m i n o u s m a tt e r and r a w fi b r e The last regulation may also be extended to the estimation of the quantity of shell present.

1. E s t i m a t i o n o f m o i s t u r e. 5 grammes of material (i. e. fine-crushed chocolate mass) are left to dry (if possible in a double-walled glycerine drying chamber) for about 6 hours at a temperature of 105 Deg. C., the loss of weight of the material being estimated as moisture. The drying should

not be continued longer than 6 hours, as fatty material is liable after the expiration of this time to recover some of its weight, owing to the oxygen of the air entering into chemical combination with the fat which rises to the surface or detaches itself from the material. When analysing chocolate, great care should be taken to prevent the mass from melting down and running together at one point. If this occurs, the following treatment must be adopted: A shallow watch-glass is filled with about 10 grammes of sand, well washed and dried, a very fine sand such as so-called sea-sand being preferable to others, the glass then transferred to the drying closet, cooled, and finally 5 grammes of the fine-crushed chocolate added. The mixture is then deposited for a period of 6 hours in the drying chamber, at a temperature of 105 Deg. as indicated above and the weight of the sand deducted when finally calculating the value of the moisture.

If as low a quantity as 5 percent of gelatine has been added to the chocolate, as much as 10 percent of water can be added without in any way affecting the appearance of the material, although such a proceeding is exceedingly detrimental to the taste and durability of the preparation. Such chocolates usually have a dull surface and, if stored in a warm place, are apt to break up and become paler in colour; this result can, however, be prevented by an extra addition of fat. Too high a[163] fat content points in any case of additions of gelatine. P. Onfroy[164] determines the addition of gelatine by boiling 5 grammes of chocolate chips in 50 cubic centimetres of water, adding 5 cubic centimetres of a solution containing 10 percent of lead acetate, and then filtering the whole. If gelatine is present in the chocolate, the liquid, on a few drops of saturated picric acid being added, leaves a yellow, amorphous sediment. If the addition of gelatine is very trifling, the gelatine is held in check or neutralised by the tannic acid. The defatting is then effected by ether and the chocolate stirred up with 100 cubic centimetres of hot water. 5-10 cubic centimetres of a solution of lye containing 10 percent of alkali and about 10 cubic centimetres of the above-mentioned lead acetate solution are added. The compound of gelatine and tannic acid is soluble in the hydrate of the alkali, and is afterwards re-deposited by the action of the lead acetate, so that it can easily be detected by means of picric acid in the neutralised filtrate. As picric acid is incapable of effecting the deposition of the theobromine, the deposition observed can only be caused by the presence of gelatine.

Like gelatine and glue, the addition of a quantity of adraganth has the power of binding the moisture and saving the fat. A method of estimating the quality of this vegetable gum, of which at the most 2 percent should be present, has recently been described by Welmans; this method is explained on page .. in the microscopic section.

2. E s t i m a t i o n   o f   a s h[165]: 5 grammes of material are heated in a platinum vessel, pan or flat tray, the latter or other similar shallow receptacle being the most suitable, holding from 25 to 30 cubic centimetres. Care should be taken when heating that the extremity of the Bunsen flame only touches the bottom of the vessel. The resulting gases are then ignited, and the c o m p l e t e l y charred mass pressed or stirred to a powder by means of a platinum wire or rod hammered flat at the end; the pan should be frequently made to revolve and its contents continually

stirred during heating, care being taken, too, to hold it slanting the whole time. The pan should be held in this way over a moderate flame until the ash assumes almost a white colour. As soon as this occurs, the pan should be cooled down and the ash uniformly saturated with a concentrated watery solution of carbonate of ammonia, whereon the vessel is placed in the drying chamber and dried at a temperature of 100 Deg. C. The contents of the pan are then heated again very cautiously over the Bunsen flame, care being taken that the bottom of the vessel is only allowed to become red-hot very gradually and to remain so for a very short time; the pan is then covered up and transferred to the dessicator to be cooled, and, on the completion of this process, its weight determined.

After repeating the saturating process with the solution of carbonate of ammonia, drying and heating for a short time as previously described, the accuracy of the weight first obtained is again tested.

3. Estimation of silicic acid in the ash When examining cocoa powders and chocolate mass, the determination of the silicic acid content of the ash is sometimes a necessity, as this facilitates the detection of any shells which may have been added.[166] The ash of the cacao bean contains only between 0·25 and 1·0 percent of silicic acid, while that of the shell shows on analysis as much as 9 percent; it must, however, be taken

into consideration that an unusually high value for silicic acid in the finished powder might be caused by impurities in the chemical or other agents used to effect the disintegration of the cacao. The signs of the presence of an extraordinary quantity of silicic acid are, according to C. R. Fresenius (Introduction to quantitative analysis)[167] a higher percentage of the ash itself than usual, and the quantity of ash used for the test should not be too small; it should further be remembered that certain cacao preparations, such as, for instance, the Dutch cocoa powders, contain large quantities of carbonic mineral matter, and the special treatment explained by Fresenius when dealing with such preparations separately should be applied.

4. Estimation of alkalis remaining in cocoa powders. The ash obtained from 5 grammes of cocoa powder is washed out of the platinum pan into an ordinary water glass or tumbler, distilled water only being used for this purpose, afterwards finely crushed with a glass rod and heated to boiling point. The liquid is then allowed to settle, filtered and re-washed. At this stage 5 cubic centimetres of n/1 sulphurous acid are added, the liquid again heated to boiling point and titrated with 2/n or n/4 alkaline lye. In this way the quantity of added carbonic mineral matter is determined, in addition to the amount of carbonate present in ordinary cocoa powders, which is formed from the organic acid minerals when the ash is

produced. Welmans has determined these values in the commonest varieties of beans and placed the results obtained at our disposal for the second edition of this book. These results are as follows:

a) Unshelled roasted beans

Per cent.	Ariba I	Ariba II	Caracas I	Caracas II
Ash	4·198	4·02	7·52	4·376
Soluble in water	1·698	1·66	1·34	1·676
Insoluble in water	2·5	2·36	6·18	2·70
Alkali (considered as potash)	0·6417	0·6417	0·596	0·9936

Per cent.	Guayaquil	Trinidad	St. Thomé
Ash	5·12	3·6	3·92
Soluble in water	2·11	1·565	1·604
Insoluble in water	3·01	2·035	2·32
Alkali (considered as potash)	0·84	1·125	0·67

b) Shelled, roasted beans:

Per cent.	Puerto Cabello	Ariba I	Aribav II	Caracas I
Ash	3·62	3·701	3·49	3·845
Soluble in water	1·72	1·423	1·315	1·76
Insoluble in water	1·90	2·273	2·175	2·08
Alkali (potash)	0·603	0·323	0·388	0·8725
Alkali in powdered cacao with 33⅓ percent of fat				

Per cent.	Caracas II	Guayaquil	Trinidad	St. Thomé
calculated	0·808	0·436	0·52	1·169
Ash, calculated as above	4·822	4·959	4·676	5·152
Ash	3·62	3·926	3·277	3·27
Soluble in water	1·62	1·476	1·727	1·34
Insoluble in water	2·00	2·45	1·55	1·93
Alkali (potash)	0·4478	0·402	0·4209	0·4048
Alkali in powdered cacao with 33⅓ percent of fat calculated	0·600	0·54	0·594	0·542
Ash, calculated as above	4·85	5·26	4·39	4·38

These tables show that:

1. The ash of cocoa powder (containing 33-1/3 percent of fat) is never more than 5·5 percent.

2. The maximum amount of alkali (calculated as potash) is 1·2 percent.

3. The ash soluble in water is always less than that insoluble in water. A reverse proportion shows a larger amount of alkali, that is, alkali has been added.

In addition to the importance of determining the amount of alkali in cocoa powder, it is very desirable that analytical chemists should agree as to the methods to be adopted, since the determinations of alkali seldom agree and may differ as much as 0·3 percent.[168] The method of calculating the results should also be defined, that is to say, an agreement should be arrived at as to whether the alkali should be

357

expressed as K_2O, K_2CO_3 or Na_2CO_3.[169]

Cacao which has been rendered miscible by means of ammonia, sometimes contains a small amount of ammonia, probably in combination with an organic acid. To detect it, the Cocoa powder should be distilled with water, which gives an alkaline distillate, as the ammonia salt would be decomposed at the temperature of boiling water. The ammonia can be volumetrically determined in the distillate with sulphuric acid.[170]

5. Determination of the Fatty Contents. In this operation 5 grammes of the finest powdered bean i. e. the finest cocoa powder (in the case of chocolate, which must be finely flaked, 10 grammes) should be mixed with an equal quantity of evenly grained quartz sand in a warmed mortar, and then transferred per filter to a Soxhlet's apparatus, wherein it can be extracted with ether for from 10 to 12 hours at a stretch. The previously weighed carboy, which now contains the fatty contents in solution, is placed on a water bath, and the ether extracted as far as possible, after which the fatty residue remaining is dried by first introducing the vessel in a water oven and afterwards allowing it to stand for 2 hours in a dessicator. The increase of weight in the flask is due to ether extract, consisting almost exclusively of fat. It is true that small proportions of theobromine will have been simultaneously dissolved (perhaps about 0·1 g.) but no special significance need be

attached to them. If it should seem advisable to avoid even this slight drawback, petroleum ether with a boiling point of 50° C. should be employed instead of the ordinary variety.

Welmans[171] has further described a quick and practical method for determining fat in cacao and its preparations, which is not only of value as a check on the extraction method, but also serves as a determination of the constituents soluble in water. It is carried out as follows:

5 grammes of Cocoa powder or cacao mass, which need not be very fine, or 10 grammes of chocolate are stirred for some minutes in a separator or cylinder with 100 ccm of ether (saturated with water) until coherent particles are no more visible, that is to say, until the factory degree of fineness has been attained. In two minutes all will have gone to powder even if the chocolate has not been rubbed down but is in pieces; 100 ccm of water (saturated with ether) are then added, and the mixture agitated until a complete emulsion takes place. With powdered cacao, especially those kinds rich in fat, that occurs in ½ to 1 minute, and with chocolate in 2 minutes. It is then allowed to rest until the emulsion separates, which at the ordinary temperature of 15-20° C. usually occurs in 6-12 hours in the case of chocolate, and 12-24 hours with cacao. The greater part of the water separates first and, usually, amounts to 90-98 ccm with chocolate and 70-86 ccm with cocoa. The powdery portion of the cocoa or chocolate floats on the surface of the aqueous layer at the bottom of the ether layer. Only husk, sand, particles of cacao beans, added starch, etc. accumulate at the bottom of the separator and are to be removed with the aqueous layer, which in the case of chocolate contains the sugar, but usually no trace of fat. The ether layer, which freely separates from the emulsion in the time mentioned, is quite clear and from 25 to 50 ccm can generally be pipetted off and an aliquot part poured into a

measuring cylinder or graduated tube, or into a 25 or 50 ccm flask. If the ether solution of fat is not sufficient in quantity, the separation can be effected after removing the aqueous liquid by twirling round the separator. The turbidity soon disappears and the non-fatty particles quickly sink to the bottom. The ether solution of fat can also be examined araometrically, as with milk fat, by Soxhlet's araometric method, after forcing it by means of an india rubber ball, into a pipette or burette, but the constants to be used in that case have not been ascertained. After the ether has been distilled off, in the normal manner, the weight obtained must be calculated for 100 ccm and a small correction made. For example, if 50 ccm of the ether solution of fat give a residue of 0·8 gramme, then 100 ccm represents 1·6 gramme. But this 1·6 gramme has not been obtained from 100 ccm of the original (water saturated) ether, but from 100-x ccm, x representing the number of cubic centimetres corresponding to 1·6 gramme of cacao butter and, as the specific gravity of cacao butter is nearly = 1; the equation becomes (100-1·6): 100 = 1·6: x; x = 160/98·4 = 1·627 gramme; so that the 5 grammes of substance would contain 1·627 gramme of fat or 32·54 percent.

The remaining aqueous solution contains the whole of the constituents of cacao or chocolate which are soluble in water. It is measured into a graduated cylinder and its volume ascertained. Then, after the entire amount has been evaporated to dryness, the residue is calculated on a percentage basis. The following procedure, however, is preferable. 10 ccm of the liquid are evaporated and the residue well dried in a vacuum before it is weighed. Multiplying the ascertained weight by 10, we obtain the amount of cacao or chocolate soluble in water and present in 5 and 10 grammes of either substance respectively. The amount of s u g a r in the aqueous extract can be determined in the following manner. 50 ccm of the extract are heated in a water bath and thus separated from ether; afterwards 2 ccm of lead acetate are added and the whole immediately transferred to a special kind of filter paper. The solution is now polarised in the usual way and the number of grammes of sugar thus ascertained converted into ccm by division (1·55 being the unit) and then the result subtracted from 100, which gives the volume of water present in 100 ccm of sugar solution, and so by further division until the percentage of sugar in chocolate is finally obtained. If the polarisation yields more sugar than the weight of the total residue, it is an indication that dextrine is present as an adulteration. The quantitative determination of dextrine, which is sometimes added to cocoa powder as well as to chocolate, for like gelatine and tragacanth it holds water together and so ensures a saving of fat, is best carried out in P. Welman's polarising method.[172]

As the amount of fat obtained from 5 grammes of a cacao preparation does not suffice for tests of purity, a larger quantity must be extracted in order to carry out the following investigations. This has reference to

1. The determination of the melting point;
2. The determination of the iodine value (Welman)[173];
3. The determination of the saponification value;
4. The determination of the acid value;
5. The determination of the Reichert-Meissl value;
6. Polen's value[174];
7. Cohn's investigation[175];
8. Melting point of the fatty acids;
9. Refraction of the fatty acids;
10. Iodine value of the fatty acids;
11. Determination of the refractive index at 40° C. in Zeiss' butter refractometer.

The following process is usually adopted in the determination of the melting point of cacao fat:

The melted fat is sucked up a glass capillary tube, the internal diameter of which does not exceed 2 mm (fluctuating between 1·8 mm and that measurement) to somewhat above the part of the tube which is graduated into tenths, and then so much of the capillary tube cut off as suffices to make the fat column there half the height of the bulb of the mercury thermometer used in the experiment.

As fresh molten fat has a very variable melting point, it is absolutely essential that the fat in this experiment be allowed to cool about a week in some dark chamber, and, because only after the expiration of this period can the melting point be designated as a constant, not to proceed with the further determination until this necessary stage has been reached.

To carry out this determination the capillary tube is attached to the bulb of the mercury thermometer by means of a rubber ring in such a manner that the column of fat occurs directly in the middle.

The whole apparatus is now hung in a test tube of 2½ cm internal diameter, which is just so far filled with water that this can only penetrate to the fat in the capillary tube which is open at both ends from the under side. To regulate the flow of heat, this test tube is further introduced into a beaker also filled with water, which is heated first. As soon as the fat is melted, the water penetrates to the capillary tube and pushes along the fat column.

The reading is now taken at once the degree registered, the thermometer showing the melting point of the fat.

We need not here launch on an exact description of the above mentioned determination, but will only stay to point out the oft-mentioned book of R. Benedikt's, entitled "Analyses of Kinds of Fat and Wax", as enlarged and issued by F. Ulzer after the death of the author (Berlin edition, J. Springer).

Should a doubt arise in comparing the results given by these six tests, which may happen with some kinds of ordinary cacao butter, the employment of Björklund's empirical ether test[175] or Filsinger's alcohol-ether test is to be recommended, which latter is carried out as follows.[177]

3 grammes of cacao butter are dissolved in 6 grammes of ether at 10° C. Should the resulting solution be clear, this is an indication that no wax is present. The solution is then introduced in its test tube into water at 0° C. and the length of the time which transpires before it begins to become cloudy or to deposit flocculent matter, observed, also the temperature when the solution again becomes clear.

If the solution becomes turbid before ten minutes have elapsed the cacao butter is not quite pure. Pure cacao butter becomes turbid in from 10 to 15 minutes at 0° C. and clear again at from 19-20° C.; an admixture of 5 percent of tallow renders the solution turbid at 19-20° C. in 8 minutes and it becomes clear again at 22° C.; 15 per cent of tallow give a turbid solution in from 4-5 minutes at 0° C. that becomes clear again at 22·5-28·5 ° C. Filsinger[178] has suggested a modification of Björklund's test. In his method 2 grammes of the fat are dissolved in a graduated tube in a mixture of 4 parts of ether (S. G. 0·725)) and 1 part of alcohol (S. G. 0·810). Pure cacao butter should remain clear after some lapse of time, whereas foreign fats and more especially tallow preparations cause a separation. But Lewkowitsch[179] maintains that this test is not be relied on, as genuine kinds of cacao butter will crystallise out from the ether alcohol solution at 9° C. and some at 12° C.

363

Yet we are nevertheless of the opinion that liquid fats are of no great moment at the present time, for they always involve a considerable lowering of the melting point and so greatly impair the fracture of the chocolate. Fats such as tallow, or the like, must be used, and these are detected both by their flavour and by Björklund's test. Adulteration is therefore very rarely met with in the German chocolate industry, thanks to these facts and the rigid self-control practised by the Association of German Chocolate Manufactures and the sharp supervision exercised by the inspectors of articles of consumption in that country. The only regularly occurring adulterations are connected with the preparation of Cocoa powder and consist in substitutions of finely ground cacao husk; the detection of which still remains most difficult and uncertain; and even here it is rather the Dutch firms which are culpable; and generally speaking it is a trick of smaller manufacturers, who consider such an admixture as quite the normal procedure.

6. Determination of Theobromine and Caffeine. Methods for the ascertainment of the quantity of theobromine are so numerous that it would be impossible here to enter into the detail of their advantages and disadvantages. Of the different processes adopted in the determination of the cacao diureide perhaps only Eminger's is worthy of consideration at present, and this is described

fully in the following paragraphs, as best corresponding to our present knowledge of the subject and its requirements, and most deserving recommendation to chemists and food analysts on account of its reliability.

For the practical testing of cacao preparations the splitting up of the diureide has no special advantage and so we can at once proceed to treat of the compound particle, though rather inclined to maintain that the diureide has very little importance on the whole, for it establishes no basis from which we can judge of the quality of the various products.

The procedure in Eminger's process is as follows:

10 grammes of powdered bean of cacao preparation are placed in a weighed glass flask, then stirred up with 100 grammes of petroleum ether and allowed to settle. The petroleum ether is next carefully poured off, without disturbing the sediment, and the treatment repeated several times. After the last decantation, the residue is well drained, then dried in the flask and weighed. The difference in weight of the residue and the former figure represents the amount of fat. An aliquot portion of the residue (about 5 grammes) is then boiled with 100 grammes of a 3-4 percent strong sulphuric acid in a flask connected with a reflux condenser, until cacao red is given as a resultant, a task which occupies three quarters of an hour. The contents of the flask are then poured into a beaker, and neutralised, whilst hot, with barium hydroxide. The whole is then mixed with sand in a basin and evaporated to dryness; afterwards the dry residue is introduced into a Soxhlet apparatus on a paper cone, and there extracted for 5 hours with 150 grammes of chloroform. The latter is carefully distilled off and the residue dried for a period of one hour at 100° C. As previously stated, the separation of the two diureides is not necessary and in commercial analyses it is sufficient to state the amount of

each separate substance after the removal of fat by means of some suitable solvent. But should the splitting up be desired, then Eminger's method should be adopted, which depends on the solubility of caffeine in carbon tetrachloride.[180] With that object, the mixture of fat, theobromine and caffeine is treated in the flask with 100 grammes of carbon tetrachloride and repeatedly agitated for one hour. After filtration, the carbon tetrachloride, which now contains fat and caffeine, is distilled off. The theobromine left undissolved in the flask and the filter used to filter the carbon tetrachloride solution are then extracted with boiling water, the solution is filtered and evaporated to dryness, the residue representing theobromine. The separation of caffeine and theobromine can also be effected by cautious treatment with caustic soda, so dissolving the theobromine and leaving the caffeine untouched in its entirety.[181] (Cf. Riederer.)

7. Determination of Starch. This can only be of importance in rarer instances, as the starch naturally present in raw cacao generally varies between 9 and 10 percent, and there is no chemical method of separating foreign matter from cacao starch. But should the necessity arise, a determination can be carried out as follows.

In order to render the starch more easily gelatinisable, the fat is first removed by treating 5 grammes of cocoa powder or 10 grammes of a cacao preparation with ether and then with an 80% solution of alcohol to separate any sugar, theobromine and cacao red. The residue is then mixed with water and subjected to a steam pressure of from three to four atmospheres, which converts the starch into a soluble body known as amylo-dextrine. This operation is generally carried out in an autoclave or strong copper vessel[182] provided with an air-tight and removable cover, the open flask, containing the sample to be gelatinised (1 part of cacao and 20 parts of water) being placed in the vessel half immersed in water.

After screwing on the lid, the temperature of the interior of the vessel is raised to 133-144° C. corresponding to a pressure of 4 atmospheres, and maintained at that pressure for three or four hours in order to allow the action to proceed on the mass for gelatinisation of the starch. The flask is then removed from the apparatus and the contents allowed to settle for a few minutes; the liquid is filtered hot, the filtrate amounting to about 250 or 300 ccm after the filter has been washed a few times with hot water. Only the cell fibre remains on the filter, whilst the starch is dissolved in the filtrate. This is now heated with 20 ccm of hydrochloric acid in a flask connected with a reflux condenser, whereby the starch is converted into dextrose. The sugar solution is neutralised with sodium carbonate, clarified with basic lead acetate, any excess of the latter being removed with sodium sulphate, finally filtered, and the whole made up to 500 ccm. The sugar is determined in this solution by titration with alkaline copper sulphate solution and from the number of cubic centimetres required for the precipitation of the red cuprous oxide, the quantity of sugar can be ascertained. As 99 parts of starch are equal to 108 parts of dextrose or grape sugar, the following calculation must be made.

$$\frac{\text{dextrose starch}}{108:99} = \left\{ \begin{array}{c} \text{dextrose} \\ \text{found} \end{array} \right\} : x$$

In the determination of sugar with copper sulphate it is more advantageous to follow up F. Allihin's[183] method, in which the cuprous oxide is reduced by hydrogen gas to metallic copper, weighed as such, and so the sugar calculated, or the cuprous oxide can be collected on an asbestos filter and weighed in that condition. The cuprous oxide must be previously washed with hot water, alcohol and ether, which must be completely removed by subsequent drying in the air bath, since an error of even 1 milligramme would seriously affect the final result. Then again, the amount of sugar may be determined by polarisation, a process which has also its own particular advantages.

The chemical determination of starch is only in a limited degree effectual in a recognition of an admixture of foreign starch in cacao preparations. If more than 10-15 percent of starch, as calculated on the crude bean, has been found, it must certainly be assumed that there is an admixture of foreign starch, but chemistry affords no assistance by which foreign starch may be separated from the genuine starch of the cacao bean. For that purpose the foreign starch must be observed under the microscope, which not only serves to detect its presence but affords a means of estimating the amount present to an approximate degree, and its characteristics. Great care should be exercised, or the result may be easily exaggerated. Standard preparations, i.e. which have a known percentage of starch constituent, prove very serviceable when comparing.

If Welman's agitation method has been used for determining the fat, the starch will be found in the sediment. The amount of foreign starch can also be determined by Posetto's[184] method, which depends on the intensity and permanency of the iodine reaction. In the latter test 2 grammes of the powdered or finely divided cacao preparation are boiled with 20 ccm water in a test tube for 2 minutes, cooled, and without disturbing the liquid, 20 ccm of water and 5 ccm of iodine solution (5 grammes of iodide and 10 grammes of potassium iodide in 100 ccm of water) are added. The liquid from genuine cacao, according to the variety used, turns brownish or light blue, changing in a short time (12 minutes at the most) to brown and red. On the other hand, chocolate or a cacao preparation adulterated with not more than 10% wheaten or potato starch, chestnut, maize or commercial dextrine, will give a blue coloration lasting for 24 hours. It must be noted that the result in Posetto's test is influenced by the amount of alkali, so that with disintegrated cacao, for instance, a considerable quantity of iodine has to be added before the blue coloration takes place, and this more especially if the potassium carbonate employed contained caustic alkali. Such preparations finally become coloured, but generally show a mixed colour (blue and yellow): green to greenish brown.

8. Determination of crude Fibre. This can be carried out in two ways; either by König's new process as employed by Filsinger for cacao or by the older method of Weender's[185] as follows:

3 grammes of the defatted and atmospherically dried substance are boiled for ½ hour with 200 ccm of a 1·24 percent solution of sulphuric acid. It is allowed to settle, then decanted, and the residue boiled twice with the same volume of water. The decanted liquids are allowed to settle in cylinders and the sediment added to the rest of the

substance, which is then boiled half an hour with 200 ccm of a 1·25 percent solution of caustic potash, filtered through a weighed filter and the residue twice boiled with 200 ccm of water. The cellulose-like substance collected on the filter is washed first with hot water, then with cold, afterwards with alcohol, and finally with ether.

After being dried and weighed, it is incinerated and the necessary corrections made for ash.

The process worked out by Henneberg is the one usually adopted for the determination of crude fibre in vegetable matter. Recently H. Suringer and B. Tollens[186] and more particularly König[187] have pointed out that in Weender's process the so-called pentosan (sugar derivative) of the composition C_5 H_{10} O_5, which comprises a not inconsiderable portion of crude fibre, would undergo a disproportionate alteration, so that the analytical results thus obtained would not represent the amount of cellulose correctly. The crude fibre must therefore be treated in such a manner as to eliminate the pentosan. König attains that result by treating 3 grammes of the defatted substance with 200 ccm of glycerine (1·23 sp. gr.) containing per litre 20 grammes of concentrated sulphuric acid, under a pressure of three atmospheres, for one hour. It is then filtered through an asbestos filter whilst hot, and after being successively washed with hot water, alcohol and ether, it is weighed, incinerated and the ash weighed. The difference between the two weighings expresses the amount of ash-free crude fibre.

Filsinger has determined by König's method the amount of crude fibre in a series of different varieties of bean, the results of which have already been given on page 72. Which process is the better has yet to be established, and in issuing results as data the method employed has always to be indicated owing to the many variations which arise.

9. The determination of cacao husk, which will be for the most part a matter of ascertaining the amount of raw or crude fibre, could formerly only be effected by means of the microscope. In 1899 Filsinger[188] proposed a method of levigation which according to P. Welman's[189] gives trustworthy results. Manifold treatises have been devoted to the subject, and it would be advisable to turn a few of these up and compare the details of the accounts.[190] In this method, which works best with the modifications suggested by Drawe (see below) 5 grammes of cocoa or chocolate are defatted with ether and dried, then ground in a mortar after a little water has been added, and levigated with about 100 ccm of water in a cylinder. The liquid is allowed to rest for some time and the suspended matter poured off almost to the sediment, which is again shaken up with fresh water, allowed to settle, and the operation repeated until all the fine particles have been floated off and the water over the sediment no longer becomes cloudy, but remains clear after the coarse and heavy particles have settled down.

The powdery sediment is collected on a watch glass, dried in the water bath, and after being cooled down in a desiccator, weighed. The weighed residue is then softened with caustic soda and glycerine and examined under the microscope. The presence of any cotyledon particles must be carefully observed, such as have escaped separation in the grinding and levigation, and whether particles of husk or epidermis or germ preponderate. With proper levigation only traces of cacao substance, especially here recognisable by the cacao starch, should be present. The sand, which always adheres to the shells in the fermenting and drying operations, is also easily recognised and many indications as to the nature of the article under investigation can be noted by the use of a simple magnifying glass applied to the washed residue on the watch glass before drying.

Examined in that way, a sample of so-called Cocoas from unshelled beans gave from 6 to 8 percent of husk; usually good cacao powder shows a maximum of 2·5% husk. It is true that from this Filsinger-Drawe procedure the correct percentage of shell can only be estimated in very rare instances, for when it is necessary to be absolutely fair to all concerned in the manufacture, the cacao must be so often washed until no grains of cacao starch are visible under the microscope; and so the result is often too small, more especially in the case of the finer qualities. But when all particles of starch have been removed, the finer particles of shell have often been taken along with them. Yet when the residue certainly exceeds the standard percentage of shell, it may be taken for granted that adulteration with husks has been carried to excess, or that the cleansing processes have not been effectively carried out. There is no other method which yields the same degree of certainty.

The result obtained by the levigation method can be controlled by the previously mentioned methods of Weender or Filsinger, as well as by the determination of any silica in the ash (page 256).

Latterly the admixture of cacao husk with the cheaper kinds of cocoa powder has largely increased, therefore the determination of the amount of husk in cacao preparations has become of special importance.

10. Determination of sugar. There are three methods for the quick determination of sugar, two of them polarimetric and the third consisting of taking the specific gravity of the solution obtained by shaking up the cacao with water. It is as well to note that in all these methods the result includes the normal amount of sugar in cacao, which Welmans[191] gives at 0·75-2 in cocoa and 0·4-1·0 percent in chocolate. That source of error is of no special significance, for, as Welmans has shown, it is compensated for in the course of the succeeding operations, so that these methods are of service.

For official investigations under this head the statutes of May 31st 1891 and May 27th 1896 respectively together with the instructions issued by the council concerning the carrying out of the process (Berlin, July 9th 1896, and Nov. 8th 1897, E) constitute a standard.

They read as follows: "Half the normal weight (13·024 g) of chocolate is damped with alcohol and then warmed for 15 minutes with 30 ccm of water on the water bath. While still hot, it is poured on to a wet filter, the residue again treated with hot water, and until the filtrate nearly amounts to 100 ccm. The filtrate is to be mixed with 5 ccm of basic lead acetate solution, allowed to stand for a quarter of an hour, then clarified with alum and a little alumina, made up to a definite volume (110 ccm) and polarised." But it is to be noted that these instructions are not exhaustive enough, and prove particularly deficient as regards the employment of water, also through their non-observation of the errors which can arise in using basic lead acetate, though it is true that these are only of a minor character.

The Berlin chemist Jeserich (ex officio) had a rather hot dispute with the official over the matter, who declared that his results were false in spite of all protest, until he finally proved that it was not these results but the process advised by law which lacked correctness. He described the rencontre in very lucid if drastic detail to an assembly of official chemists.

Something similar happened to the present editor, who in his office of sworn chemist was called upon to determine the amount of sugar and starch present in certain crumb chocolates on the one hand, and the amount of cacao material on the other. As the official inspectors insist on their prescriptions being carried out with scrupulous exactitude, he found it necessary to give a double result, the one in accordance with these prescriptions, and the other when double the amount of water was used, taking care to explain the whole matter at length. But it occasioned some surprise, and finally the task of investigating and testing was withdrawn and given to another.

Another polarimetric method, recommended by Woy[192], is carried out as follows. Two portions of half the normal weight (13·024 grammes) of rasped or shaved chocolate are placed in 100 ccm and 200 ccm flasks respectively, moistened with alcohol, then treated with hot water and stirred up till the sugar is dissolved. 4 ccm of basic lead acetate solution are added to each flask, by which means the chocolate in suspension loses its viscosity. After being cooled, the solutions are made up to the marks, well mixed and filtered. Two quickly filtering liquids are thus obtained, which are then polarised in 200 mm tubes. With chocolate containing meal, the temperature must not exceed 50° C. From the two polarisations, the following equation results: a (100-x) =

b(200-x), in which a and b are the results of polarising, and x the volume of the insoluble substances, including the lead precipitate, contained in the half normal weight. The product of the equation gives the amount of sugar present. Woy's method has the great advantage of avoiding the error due to the volume of the undissolved cacao and lead precipitate.

The third method, as adopted by Zipperer[193], is as follows: 50 grammes of chocolate, finely divided with an iron grater or rasp, are treated with exactly 200 ccm of cold water, frequently stirred for 4 hours, then poured on to a previously moistened and well wrung pointed bag. The specific gravity of the filtrate is taken in an araeometer, specially constructed for the purpose by Greiner of Munich on lines suggested by Zipperer himself. On the scale of the araeometer is given the percentage amount of sugar in the chocolate, from 5 to 5 percent, with subdivisions of one percent, so that the reading can be quickly taken, without correction.

In the determination of sugar by weight, the chocolate is first defatted with ether, the sugar extracted with alcohol, then inverted, the inverted solution treated with Fehling's solution and the copper precipitate weighed. The process has little to recommend it, being troublesome and admitting of a large margin of errors.

Here again much has been written of late[194] concerning the two former methods, their liabilities to error and the avoidance of these, yet without bringing to light anything which calls for a specially detailed treatment in this book.

11. Determination of Albuminates. The determination of albumin is frequently required in the analysis of cacao powder and is necessary to the ascertainment of its nutritive value. The determination of nitrogen is determined by mixing 0·5 grammes of finely powdered bean with soda lime and burning the mixture in a tube. (This determination of nitrogen is a necessary part of the process.) Thus ammonia is formed, which is passed through a known quantity of sulphuric acid. When the combustion is finished, the acid solution is titrated with a standard solution of barium hydroxide, and from the quantity consumed the percentage of nitrogen is calculated. But as the diureides also contain nitrogen (31·1 % of the theobromine and caffeine present) the nitrogen corresponding to this amount must be deducted from the total quantity of nitrogen yielded by combustion and the remainder multiplied by 6·25 will indicate the amount of albumen present as a constituent.

Another and better method of determining the nitrogen is by Kjeldahl's[195] process. It has been frequently subjected to modifications, but was originally carried out as follows. 0·25 grammes of the nitrogenous substance (cacao preparation) is heated on the sand bath together with 20 ccm of concentrated sulphuric acid and a little quicksilver, till the solution becomes colourless or only of a very pale yellow. After diluting with about 200 ccm of water, it is made alkaline by the addition of soda lye (which must of course be entirely free from nitrogen, the same remark applying to the sulphuric acid used) and, potassium or sodium sulphide being added, it is then distilled, and the ammonia given off collected and determined as above described. As this method also determines the total amount of nitrogen, an allowance must be made for the nitrogen in the theobromine and caffeine before multiplying the result by 6·25. This modification is still to be recommended as the best and most reliable.

In rare cases an excessive amount of albumen may be due to the admixture of earth-nut cake or gelatine. As to the detection of the latter adulteration, see page 254. Bileryst[196] says that earth-nut cake can be recognised by its high percentage of albumen content, amounting to between 45 and 47 percent.

12. Investigation of Milk and Cream Chocolate. The tests bearing on these products really constitute a chapter in themselves, which has acquired special importance owing to the great popularity they enjoy and the consequently greatly increased production. According to the unanimous opinion of the Association of German Chocolate Manufacturers and the Free Union of German Food Chemists, expressed when considering the respective claims of such chocolates, it is chiefly if not exclusively a matter of determining the percentage of milk or cream, which ought not to be below 12·5 or 10%, always supposing the milk or cream to be a substitute for sugar, and this means therefore that the quantity of cacao material in the chocolate product should on no account sink below

32%. (Cf. p. 283 No. 3. Abs. 5.) The method employed in the investigation is generally the same as that suggested by Laxa in his treatise on "Milk Chocolates"[197] although it has been considerably improved by Baier and his colleagues.[198] It is here a matter of working backwards from the determination of the fatty and nitrogenous components (or caseine) to the amount of milk or cream in the chocolate. This presents a certain amount of difficultly as it is not only necessary to determine the milk, but also to establish that neither skimmed or whipped material (either in part or entirely) has been employed. Yet it is possible here to proceed with absolute certainty, as Baier[199] convincingly demonstrates, by taking into consideration the relative proportion of milk fat, called caseum or caseine.

If it is desired finally to characterise the respective chocolates, determinations of the q u a n t i t y of milk fat present and the amount of milk product used become essential. Baier gives both as calculable (cf. footnote 1)[200], the Reichert-Meissl number of the total fat being ascertained, and from this, subtracting the R.-M. number of the cacao fat present[201] the quantity of milk fat, finally the amount of caseine, milk sugar, mineral matter and other factors. No details of this somewhat extensive calculation are proved in the original.[202] We give the following regulations (Laxa-Baier) for carrying out the determination of the caseine, together with the necessary formula.

20 grammes of fine divided chocolate are loosely introduced into a Soxhlet's extracting apparatus, and there extracted with ether for a period of 16 hours. Of the residue, 10 grammes are used for testing in connection with caseine, and this after the ether has evaporated. These are mixed up in mortar with gradual and even addition of a 1% solution of sodium oxalate, so that no lump formations occur, and

then brought into a marked carboy of 250 ccm capacity, until 200 ccm of the sodium oxalate solution have been used. The carboy is then provided with an asbestos net, and heated by means of a flame from the under side, until its contents are brought to boil. The mouth of the carboy is covered with a small funnel which has been hermetically sealed at its narrower end. Then boiling oxalate solution is poured into the vessel up the bend, and it is then allowed to stand over till another day, shaking however being often repeated, then filled with sodium oxalate solution up to the mark, agitated with a regular motion, and then filtered through an ordinary filter. To 100 ccm of this solution 5 ccm of an uranous acetate solution (5% strong) and drop by drop and with repeated stirring a 30% solution of acetic are added until there is a deposit. (This will require from 30 to 120 drops, according to the amount of caseine present.) Then an extra 5 drops of acetic acid can be added. This causes the deposit to stand out clearly from the liquid matter and it can be readily separated by centrifugalising. Afterwards it can be washed out with 100 ccm of solution, of which 5 ccm are uranous acetate and 3 ccm acetic acid 3 % strong, until the sodium oxalate can no longer be seen on adding calcium chloride (i. e. after about three repeated centrifugalisations). The contents of the tube are then rinsed on to the small filter by means of the wash fluid, stirred in a Kjeldahl carboy with concentrated sulphuric acid and copper oxide, and the quantity of nitrogen found converted into caseine by multiplying with the factor k = 6·37. — Bearing in mind the quantity of fat, the percentage of caseine in the original chocolate is calculated.[203]

In the following:

b = signifies the total of fatty content of the chocolate[204],
    a = the Reichert-Meissl number of the total fat,
and K = the amount of caseine as established by the Laxa-

Baier method (nitrogen contents times 6·37).

1. $F = (a-1)b/27$

Further: 1. The desired quantity of fat equal is to the R.M.N. a, of the total fat less that of the cacao fat (1·0) multiplied by the total amount of fat and divided by the average R.M.N. for butter fats = 27.

2. a) $E = 1·11 K$

b) $M = (1·11 K - 132)/100$

c) $A = (1·11 K - 21.4)/100$

2. The total amount of albumen E is equal to the amount of caseine K multiplied by 1·11, as this constitutes about 90% of the former; and as the albumen E, milk sugar M and the mineral constituents A (Ash V) are present in the milk in the proportion of 100 : 132 : 21·4, this yields the formula given in b & c.

3. $T = F + E + M + A$

3. The total quantity of milk stuff T is equal to the total of fatty contents, albumen, milk, sugar and ash.

4. $x = Q \cdot k$

$Q = F/K$

$k = const.$

In the case of milk:

$k_1 = 3·15$

$k_2 = 3·05$

$k_3 = 2·7$

$k_4 = 2·5$

4. The fatty constituency of the original milk or cream to be calculated from the formula x = Q times k, where Q is the quotient resulting when the amount of fat F is divided by that of caseine K, and k the normal caseine consistency of average milk preparation. Or it varies as the numbers k_1 etc. indicate in the case of 10% cream and so forth. Higher percentages than those given do not come into consideration.

c. Microscopic-botanical investigation.

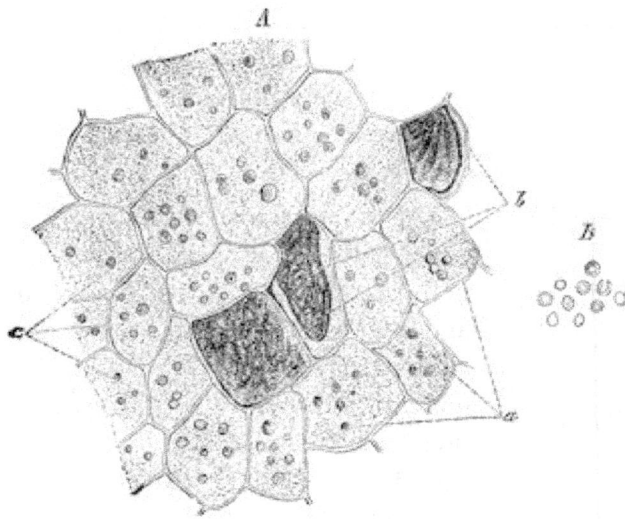

Fig. 100.

A. Parenchyma of the cotyledon after removal of fat and treatment with Iodine chloral hydrate, a: parenchyma cells with starch, b: with cacao red.

B. Aleuron particles with globois (Molisch) from parenchyma cells.

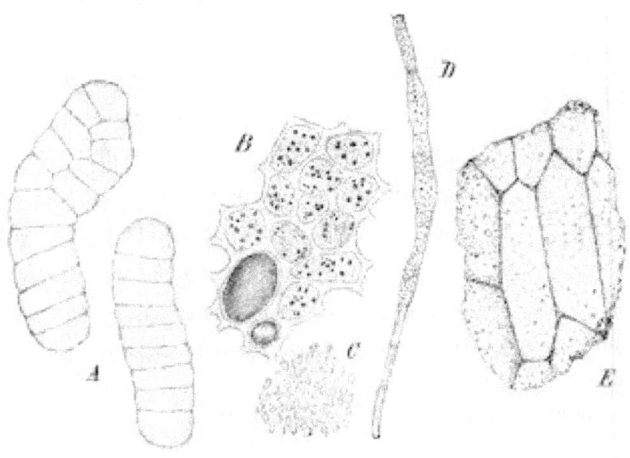

Fig. 101.

A. Mitscherlich particles.

B. Seed cells, above with starch bodies, underneath with violet colouring matter (cacao red) lying in chloral.

C. Series of yeast germs.

D. Threads of extraneous growth.

E. Epidermis and layer of cells occurring on the outer shell (enlarged 340 times).

Cacao is to no great extent particularly characterised anatomically. The parenchyma cells fig. 100 are chiefly to be noticed, containing either fat, albumin (protoplasm) aleuron granules, pigment, or cacao starch. The s t a r c h, as already remarked, consists of especially small globular granules, mostly separate, but also two or three adherent. It is somewhat more difficult to gelatinise than other kinds of starch, and it is coloured blue by iodine somewhat more slowly than many other starch granules, especially in the preparations containing fat. Cacao preparations which have been disintegrated by fixed alcalis, differ in this respect; according to Welmans, iodine first forms colourless iodine compounds, and not until the alkali has been saturated, is the blue colour developed. In such cases, care must be taken, that an excess of iodine is present. In estimating the amount of foreign starch, great care must be taken that the conspicuous bluish-black granules of the foreign starch, which immediately strike the eye, are not over estimated, which may easily occur. For control observations, mixtures containing various known amounts of starch should be tried comparatively. The pigment cells and the epidermis with the Mitscherlich's particles (figs 101 and 102) should be noticed as well as the characteristic globoids, which occur in the ash of the cotyledon tissues (compare page 67). The o u t s i d e   s h e l l more or less woody according to the origin of the bean, consists of four layers of cells; this is best recognised by the large cells of the principal tissue, which are distinguished by their form as well as by their thickened side walls from the tissue of the cotyledon. Another characteristic of this layer consists of the large number of coarse spiral vessels, which exceed those of the seed lobes in size, and finally, the inner elements of the stone cell layer, which, however, on account of their limited development are seldom to be discovered. The smooth, fine brown coloured, and light refracting fragments, which frequently appear quite structureless and have their fibrous character made perceptible only after treatment with caustic alkali, must be regarded as characteristic of the inner part of the husk or the seed membrane. The best observing medium is a solution of chloral hydrate or almond oil, as well as dilute sulphuric acid and glycerine.[205] The substance is always to be defatted with ether, before the microscopical examination. A complete extraction of the fat, according to Welmans, can occur only with exceedingly thin cuttings, in which every cell of the section would be operated on, or in powdered preparations, when the cells have been completely torn asunder by mechanical pulverisation. The fat is not extracted by solvents from intact cells, as the cell walls are impermeable by them.[206]

The detection of f o r e i g n   s t a r c h is possible only by use of the microscope; by means of standard preparations an approximate estimate may be made as to the amount and kind of meal added.[207] The examination of starch is especially facilitated by H. Leffmann and W. Beam's[208] centrifugal method: the sample suspended in water is subjected to rotation for a short time in the centrifugal apparatus. The presence of foreign starch is shown by a white layer in the resulting sediment. This layer can be collected and microscopically examined for foreign starch and husk. In the case of cacao preparations, it is always well to distinguish between unimportant traces and quantities that justify objection.[209]

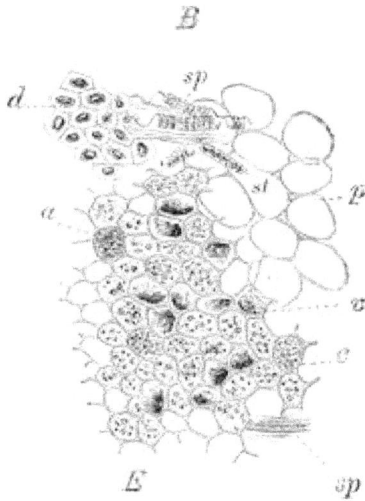

Fig. 102.

A. Silver membrane with the hairs (Mitscherlich particles) *tr*, and the crystals *f* and *K*.

B. Cocoa powder: *c* Cotyledon tissue with cells of fat and colouring matter, *p* shell parenchyma, *sp* speriods, *d* layer of dry cells.

A means of detecting t r a g a c a n t h in cacao preparations, has lately been described by Welmans[210]. 5 grammes of the cacao preparation are to be mixed with sufficient dilute sulphuric acid (1: 3) to form a thick pulp, then with 10 drops of solution of iodine (in potassium iodide) and some glycerine. A portion of the mixture is examined under the microscope (enlarged 160 times). The entire field of view now appears to be thickly sown with countless blue dots, some globular, others irregular, among which are especially to be noticed the large tragacanth cells, resembling potato starch, which are not seen in cocoa powder that is free from tragacanth, when similarly prepared as an object; the small blue dots, due to cacao starch, are visible only in the densely occupied portions.

An admixture of the c a r o b, which has been seldom

385

observed, can be easily recognised under the microscope by the characteristic reddish wrinkled tubes of the fruit pulp, which are also coloured violet by treatment with a warm solution of caustic potash.

The presence of e a r t h - n u t   o r   e a r t h - n u t   c a k e can be detected by the aid of the microscope on treatment with chloral hydrate, by the characteristic saw toothed epidermis cells of the husk of arachis seed.

H a z e l n u t   a n d   w a l n u t   p u l p so far as they are to be met with in cacao preparations, can be distinguished under the microscope by shreds of the tissue of the seed husks, in which broad streaks of spiral vessels, lying close on one another, are distinctly prominent. If in addition the woody fruit shell be admixed, it can be detected by the great number of cells.

B. Definitions of Cacao Preparations.

The following formulae have been compiled by the Association of German Chocolate Manufacturers for the purpose of fixing the definition of cacao, and we may say that we agree with same in the main, as they satisfy all just claims, and keep pace with the progress made in consequence of the introduction of the modern machinery now in use, both from a scientific and practical point of view. Only in a few points are we of different opinion, and have referred to such clearly in their place.

a) Regulations of the Association of German Chocolate Manufacturers relating to the Trade in Cacao Preparations (cocoa, chocolate and chocolate goods).
(Revision of September 16th 1907.)

I.

1. Cacao mass is the product obtained by simply grinding and moulding roasted and shelled cacao beans and no substance handled under this name may contain any admixtures of foreign matter.

2. Disintegrated cacao mass is cacao which has been treated either with alkalis, alkaline earth, or steam.

3. Cocoa powder, freed of oil (also soluble, disintegrated cacao) is the resulting product when the cacao bean is decorticated, roasted and more or less freed from its oil or also disintegrated in powder form. Cocoa powder, cacao from which oil has been extracted, disintegrated and soluble cacao may on no account

contain foreign ingredient other than an addition of roots and spices.

In the case of cacaos disintegrated with alkalis or alkali earths, not more than 3 % of alkali or alkali earth may be used in the process; they may not contain more than 8 % of ash, reckoned on cacao material with 56 % of cacao butter.

4. C h o c o l a t e . The designation "Chocolate" may only be applied to those confections which are prepared by the addition of cacao butter, vanilla, vanillin, cinnamon, cloves or other spices to roasted and shelled beans or to a disintegrated mixture of cacao and sugar.

The percentage amount of sugar may not exceed 70, and the occasional addition of other substances (medicinal, meals, and the like) is admissible, but the total percentage of these and the sugar may not exceed 70.[211]

5. F o o d   c h o c o l a t e ş   c h o c o l a t e s   f o r   i m m e d i a t e   c o n s u m p t i o ņ   and   d e s s e r t   c h o c o l a t e s . For these confections the same principles hold good, with the exception that here additions of nuts, almonds and milk stuffs are permissible, up to a percentage not exceeding 5 in total, without any declaration of the goods being necessary.

6. C h o c o l a t e   p o w d e r is a mixture of cacao material which may be disintegrated and more or less freed from oil, with an amount of sugar not exceeding 70% at the most. Spices as in the case of chocolate.[211]

7. Cacao butter is the fat obtained from the decorticated bean or cacao material.

II.

The following are especially to be regarded as adulterations of the goods mentioned under I. from 1 to 7.

1. Foreign fats;

2. Shells and other waste cacao products (dust or seed);

3. Meal, though this is not expressly given;

4. Colouring matter; the colouring of the surface of figures is permissible;

5. So-called fat economisers, such as adraganth, gelatine, and dextrine.

An addition of substances for medicinal or dietetic purposes is permissible, though in such cases the goods must be declared. The addition of any fats other than cacao butter (i. e. of any foreign fat) or of shells or waste products to cacao or chocolate or to cacao or chocolate goods is also not permissible even when these are designated in such a manner that the words chocolate and cacao do not occur in their description.

III.
Declaration of Added Ingredients.

The declaration must be transcribed in legible script and form, as e. g. "Meal" so as to be readily understood by all, and composed in German.

The declaration must occur together with the description

of contents and as part of the same on despatching original packages in retail transactions.

In wholesale trade the declaration must occur on all offers, quotations, bills and all boxes, etc. provided with description of contents.

When offered for sale or exhibited in an unpacked condition, every box etc. containing the goods must have such a declaration introduced so as to be visible to every buyer in the premises, where possible; or the declaration shall be placed on the goods themselves.

1. Skimmed milk chocolates must be literally described as such, and must be manufactured with at least 10 % of skimmed milk powder or the corresponding quantity of skimmed milk proper. Addition of ordinary milk or its powder is permissible and need not be declared;

2. Milk chocolate must be manufactured with ordinary milk containing at least 3 % of fat, and in such a manner that at least 10 % of milk powder or the corresponding quantity of milk proper are employed;

3. Cream chocolate must be prepared from cream containing at least 10 % of fat, and in such a manner that at least 10 % of a cream powder or the corresponding amount of cream, in each case containing 50 % of milk fat, are employed. It may be varied to taste with milk proper or its powder, without any further declaration being necessary.

These percentages represent a minimum. It remains at the manufacturer's choice whether he shall employ larger quantities of cream or milk.

The associated firms are further recommended to annex

the following guarantees:

a) that the powder of milk proper contain at least 26 % of fat and be prepared from a milk guaranteed as pure;

b) that the cream powder contain at least 40 % of milk or be prepared from cream containing at least 10% of fat.

It is especially emphasised that these quantities are minimums, and every manufacturer is free to add as much cream or milk as he pleases.

We particularly recommend the procuring of a guarantee from the milk purveyor as to its purity for every delivery in order to be covered against fines in case the product should prove to contain an insufficient amount of fat. Analytical testings of trial samples are also to be recommended.

By way of comparison we refer to the "Principles for Estimating Cacao Products and their Food Value" determined by the Free Union of German Food Chemists in their 8th annual assembly at Heidelberg (1909) and finally established in their 10th held at Dresden (1911), which are said to have found general acceptance from the 1st July, 1912.

b) Final Wording of the Principles of the Free Union of German Food Chemists for the estimation of the Value of Cocoa and Cacao Preparations.

I.

C a c a o   m a s s is the product which is purely and simply obtained from the roasted and shelled cacao bean by

391

grinding and moulding.

Cacao mass may not contain any kind of foreign substance. Traces of shell may only be present in minor quantity. The waste product falling in the cleansing of the bean must not be added to the cacao mass, nor may it be worked up into cacao material separate and apart from other cacao.

Cacao mass shows 2·5-5% of ash and contains 52-58 % of fat.

Disintegrated cacao is such material as is treated with alkalis or alkaline earths, ammonia or its salts, under pressure of steam.

II.

C o c o a   p o w d e r, cacao that has been pressed and its oil removed, soluble Cocoa and disintegrated cacao are synonyms for cacao mass which has been reduced to powder form after they have been partially separated from fat by expression under heat; and generally treatment with alkalis or their carbonates, alkaline earths, ammonia, and ammonia salts under a strong steam pressure are presupposed.

Cocoa powder containing under 20% of fat, as well as that treated with spices (aromatised or scented) must be declared accordingly.

Cocoa powder may not contain any kind of foreign substance. Traces of shell may only be present in minor quantity. The waste product falling in the cleansing of the bean may neither be added to the cocoa powder nor itself worked up into such a powder.

The added alkali or alkaline earths may not exceed 3 % of the raw material.

Only powdered cacao and cocoa powder which has been treated with ammonia and its salts under strong steam pressure shows from 3 to 5 % of ash on cacao mass containing 55 % of fat.

Cocoa powders disintegrated with alkalis and alkaline earths must not show more than 8 % of ash on cacao containing 55 % of fat.

The percentage of water must not rise above 9.

III.

C h o c o l a t e is a mixture of cacao material with beetroot or cane sugar and a proportionate admixture of spices (vanilla, vanillin, cinnamon, cloves and so forth). Many chocolates contain apart from that an addition of cacao butter.

The percentage of sugar may not amount to more than 68.

Addition of substances for dietetic and medicinal purposes is permissible, and then the total of sugar and such addition must not exceed 68% of the whole.

Apart from the addition of spices no other vegetable admixtures are permissible. Nor may chocolate contain any foreign fat or foreign mineral constituents. Cacao shells may only be present in faint traces. The waste product falling in the cleansing of the bean must not be added to the cacao mass, nor may it be worked up into cacao material itself.

Chocolates which contain meal, almonds, walnuts, hazelnuts and milk stuffs must be provided with a declaration indicating such addition precisely, and here again the total addition of foreign ingredients shall not exceed 68 %.[212]

The percentage of ash constituent shall not exceed 2.5.

IV.

Covering or coating material must satisfy the requirements holding good for chocolate even when the coated goods bear declarations in which the words cacao or chocolate do not expressly occur, although admixtures of nuts, almonds and milk stuffs not exceeding a total of 5% may be made without declaration.

V.

Chocolate powder may not contain more than 68 % of sugar.

VI.

Cacao butter is the fat obtained from the hulled bean or cacao mass.

Milk and Cream Chocolates.

1. Cream, milk and skimmed milk-chocolates are products which are manufactured with addition of cream, milk (skimmed or unskimmed) in a natural, thickened or dry form. They must be declared as cream, milk or skimmed milk chocolates.

2. The fat content of full milk should amount to at least 3 per cent., and that of cream itself 10 percent. If the full milk

or cream is added in a condensed or dried state, these ingredients must be in corresponding proportions. As it is at present not possible to produce a cream powder containing at least 55 percent of fat, the normal preparation of this class is, for the time being, represented by a production containing 5.5 percent of milk fat in the form of cream and milk.

3. Milk chocolate prepared from skimmed milk must contain at least 12.5 percent of dried milk or skim-milk, and "Cream" chocolates not less than 10 percent of cream or full-cream powder.

4. The percentage of the milk or cream preparation added must in all chocolates be deducted only from the percentage of the sugar, i. e. the cacao content of all chocolates containing these ingredients must be the same as in the case of the commoner varieties.

Special notice. In the case of butter chocolates, in which the cream is replaced by pure cacao fat, the same regulations naturally obtain; thus the amount of butter added must be not less than 5.5 percent of the whole, and the butter should be used in place of the sugar only.

(Regulations relating to the manner of examining chocolates as to the presence of the prescribed quantities of the above ingredients will probably be issued in the course of a year or two.)

c. Vienna Regulations.

The Assembly of Microscopical and Food Chemists in Vienna, held on the 12th-13th October 1897, the object of which was to fix a "Codex Alimentarius Austriacus", also arrived at some just and appreciable definitions, which are well worthy of repetition here:[213]

1. Chocolate should consist of a mixture of cacao, Austrian sugar capable of fermentation, further an addition of spices (cinnamon, cloves, vanilla or vanillin) amounting to as much as 1 percent of the whole.

2. Cacao mass should consist of the roasted and shelled cacao bean, ground and moulded, only.

3. Cocoa Powder should be a preparation obtained from cacao mass only by the partial expression of the 50 percent of fat which the latter contains and frequently treated with alkalis. The alkalis may reach 2 percent of the whole, and the object of the treatment with them is to effect the disintegration of the tissues of cacao or to render the cacao "soluble

d) International Definitions.

An International Congress of Chocolate and Cocoa Manufacturers was finally held in Berne on August 21st-23rd 1911, which, unlike the meeting held by the White Cross in Geneva (1908), the object of which was the prevention of food adulterations, was really international and attended by numerous manufacturers from Belgium, Germany, England, France, Holland, Italy, Mexico, Austria, Hungary, Russia and Switzerland, the total number of visitors amounting to 250.[214]

1. Cacao Mass.

§ 1. Cacao mass is obtained by roasting or drying[215] cacao beans which have previously been well cleaned and freed from the shells and dust. Cacao mass can either be disintegrated, i. e. "soluble" or untreated with disintegrating agents, i. e. "insoluble

Cacao which has been treated according to § 5 is in the real and business sense of the term to be regarded as a p u r e article of food, seeing that the treatment with alkaline carbonates or pure alkali is a purely chemical, or technical, operation. Such cacao may therefore be justly termed "pure

§ 2. Cacao mass may contain a quantity of added cacao butter proportionate to the prescribed, or suitable, fat content of the cacao preparation to be made.

2. Cocoa Powder.

§ 3. Cocoa powder should consist of defatted, or fatty, pulverised cacao mass.

§ 4. Cocoa powder which has been opened up by means of alkalis or otherwise is termed "soluble" or disintegrated cacao.

Disintegrated cacao which has been treated as described under § 5 may in the real and business sense of the word be regarded as a "pure" article of food, as the treatment with alkaline carbonates or pure alkalis is a purely chemical, or technical, operation. Such cacao may, therefore, be justly termed "pure

§ 5. The quantity of alkali used to effect the treatment described should not exceed $5·75^{216}$ grammes of potash or the equivalent of another alkaline carbonate, to 100 grammes of dry defatted cacao.

III. Cacao Butter.

§ 6. Cacao butter consists of the fat obtained from either untreated or disintegrated cacao.

IV. Chocolate and Chocolate Powder.

§ 7. Chocolate is a mixture of cacao mass and sugar, with or without the addition of cacao butter. On pulverising chocolate, chocolate powder is obtained.

§ 8. Both chocolate and chocolate powder may, if the methods of manufacture require it, be prepared from partially defatted cacao mass.

§ 9. The amount of cacao mass and cacao butter contained in chocolates and chocolate powders should be at least 32 percent[217] of the whole.

V. Milk Chocolate.

§ 10. Milk chocolate should consist of a mixture of cacao mass, cacao butter, sugar and milk or milk powder. The quantity of cacao mass and cacao butter contained in such preparations should amount together to at least 25 percent[218] of the whole.

§ 11. All chocolates which are brought on to the market under the name of milk chocolate, must contain at least 12·5 percent of milk or milk powder.

§ 12. No milk used for the preparation of milk chocolate may contain any preserving agent.

VI. Covering Matter.

§ 13. The definitions of chocolate proper apply also to

covering material.

§ 14. Covering chocolate may, without special designation, contain up to 5 percent of its weight as sold of almonds, nuts, milk or milk powder. All other additions must be clearly declared on the packages in which the covering material is sold, or in the invoices referring to it.

VII. Flavouring matter(Spices).

§ 15. All material (spices etc.) used for flavouring cacao preparations must be harmless.[219]

Name of cacao preparation	Adulteration	Mode of Detection	Reference
Chocolate		a) Microscopically	277
Cacao mass		b) By excess in glucose	264
	Meal (kind not stated)		
Coated Goods Covering		c) By decreasing the amount of ash	—
Chocolats fondants			
	Cacao husks and sawdust	a) By increasing the amount of ash and the amount of silicic acid in the ash	256
		b) Method of levigation	267
		c) Determination of fibre	266
		d) Microscopically	275

	Foreign fats and oils	a) Melting point	260
		b) Iodine value	—
		c) Saponification value	—
		d) Refractometer test	—
		e) Björklund's test	261
	Bad (rancid) cacao-butter	a) Acid value	260
		b) Reichert-Meissl number	—
Forbidden colouring matters	Yellow ochre		—
	Red ferric oxide	Increase in the amount of ash	
	Brickdust		
	Coal		
only observed in soup powders;	Cacaolol		—
only used to imitate the ash of chocolate cigars;	Zinc white and heavy spar	Analytically in the ash	—
	Besides inorganic weighting material	Increase in the amount of ash	—
	Sand Clay		
	Dextrine	Polarisation by Welmans' process	258
		a) Polarimetric test	269
	Excess of		

400

	sugar	b) Aräometric test	270
		c) Decrease in ash	—
	Excess of cacao butter	Determination of amount of fat	258
	Excess of water	Determination of moisture	254
	Gelatine	Picric acid test and albumin determinatio	—
	Tragacanth	Microscopically	277
	Earth-nut	Microscopically	278
	Earth-nut-cake	Determination of albumin	271
	Walnut- and hazelnut pulps	Microscopically	278
Cocoa-Powder	Husk		
	Foreign fat	As with chocolate	—
	Meal		

C. Adulterations of Cacao wares and their Recognition.

a) Introductory.

Cacao preparations are subject to manifold and various kinds of adulteration. The following table gives a list of proved adulterating agents, and contains in the last column but one hints as to how such foreign additions can be detected, which hints are given in more detail on various pages in this edition, the numbers of which are annexed in the last column.

Bases for the judgment of cacao preparations appear on the one hand in the definitions and formulas previously given, and on the other in the rougher and finer adulterations which we had the opportunity of detecting. We give these bases once more, at least such as we deem necessary to a proper estimation of the purity of cacao goods, and in general rather incline to the principles which Filsinger has worked out for the Imperial Health Office (Germany) and which received a hearty reception at the hands of the various unions connected with the trade.

b) The Principles.

Chocolate, Cacao material and cocoa powder (defatted and disintegrated cacao) may on no account contain any kind of foreign vegetable mixtures like starch, meal, peanut cake, hazel nut and walnut admixtures, nor cacao shells nor yet waste products, neither may it contain any mineral stuffs or foreign fats. Chocolates with meal addition must contain on the wrapper a concise and definite declaration of such addition on the wrapper. The presence of cacao shells is detrimental to the nutritious value of the cacao preparation, being little suited for human consumption, as they contain a large quantity of woody substance, and apart from this, always occur with adhering sand and earth. The removal of such shells is since the perfecting of the cleaning machinery intended for the purpose, become a very easy matter, and so none but very inferior quantities are permissible. Any additional shells (even when declared, and very fine ground) are illegal. The addition of spices or their corresponding ethereal oils are allowed, and as such may be considered almonds and nuts, more especially in the case of coating material and so forth, although they are subject to compulsory declaration.

The same conditions prevail in the case of c h o c o l a t e e n a m e l l i n g and c o a t i n g material as for ordinary chocolate, and in particular they must be free from all kind of foreign fats and cacao shells.

The use of dyes (earth-and tar-colouring matter) which are intended as substitutes for a percentage of cacao, and not merely as ornamental, is not permissible; and such dyes as are objectionable from a hygienical standpoint are impossible, even when they are used for decorating purposes. Cacao material contains on an average from 3-4 % of ash and from 50-55 % of fat.

Admixtures of glue, tragacanth or dextrine are not permissible, when they are intended to conceal an addition of water or to save the use of expensive cacao fat.

C o c o a p o w d e r contains arbitrary quantities of fat, and shows accordingly a varying quantity of ash to correspond with the amount of fat expressed. It is therefore necessary to declare the quantity of fat contained in quite a general manner and something in the following grade: skimmed milk cacao under 25 % of fat, cacao freed from oil, fatty and ordinary milk chocolate up to a percentage not exceeding 35.[220] For the same reason it is necessary to convert the established ash contents, possibly of cacao material with 50% of fat, or none at all. It is most to the purpose to convert in the case of dry material which has been freed of fat, as occasionally considerable amounts of moisture remain over from the processes of preparation. Cocoa powder which has been disintegrated without the use of potassium, sodium or magnesia agents (carbonic acid) will therefore show the same ash contents as the corresponding material freed from oil, whilst that of cacao disintegrated by means of the fixed alkalis will be greater. The ash contents of powder freed of oil may nevertheless not exceed 3 %, corresponding to a total 7 %. The mixture of cocoa powder and sugar is not permissible.

Chocolates, chocolate fondants and coating mass contain variable quantities of sugar and fat; accordingly no limits can be assigned to the ash contents of these preparations.

A unanimity of opinion as to the least possible amount of cacao for the chemical estimation of chocolate has become an urgent necessity. Hereby it should be established that in good chocolate the fatty contents, apart from the sugar,[221] exceed a definite percentage.[222] A minimum percentage of 35% of cacao mass in chocolate destined for export, which must possibly be covered, has been fixed by the council of commerce.

As percentage of chocolate in cacao the double quantity of non-fatty cacao material must be taken, on the supposition that raw cacao contains on an average 50% of fat.

c) Laws and Enactments as to Trade in Cacao Preparations.

So far traffic in cacao has only been brought under legal control in three European countries, namely Belgium, Roumania, and Switzerland. We annex in the following pages a resumé of the legal prescriptions appertaining thereto, as being of especial importance to exporting manufacturers.

1 Belgium.

The Belgian royal decree of the 18th November 1894 established on the basis of the law for articles of consumption, August 4th 1890, and article 454 to 457, 500 to 503, and 561 of the penal code book runs (according to the "Moniteur Belge" of the 3rd and 4th December, 1894, as follows:

Art. 1. It is illegal to sell, expose or hold in possession for sale, or to transmit, any other product as "all cocoa" than the fruit of the cacao tree, raw and prepared by roasting, hulling and grinding with or without addition of spices, and finally moulded into tablets or reduced to powder form.

It is permissible to sell, expose or have in possession for sale, or to transmit such cacao as has suffered a loss of butter by expressing, provided that the amount of this ingredient is not diminished by more than 20 % of the whole, under the designation "cocoa or cocoa powder"; and again under the designation "alkalinised cacao" (cacao alcalinisé) such as has had its alkali content increased in special treatment by not more than 3% of the total weight. The declaration "alkalinised" is not, if a matter of mere possession or transmittance in export, to be considered as necessary.

Cacao which has been prepared other than as above described may only be sold, exposed or held in possession for sale, or transmitted, under a special label which declares this special manner of preparation next to the word "cacao" or under a label that does not contain the word "cacao" at all.

The word "alkalinised" or any other words which indicate alterations or additions in the natural composition of the cacao must be introduced on the label in distinct and similar type to the word "cacao

Cacao in which the proportion of alkali amounts to more than 3% is regarded as injurious, and the sale, having and holding in possession or despatch of same for sale is illegal.

Art. 2. It is illegal to sell, have in possession or expose for sale, or to transmit any product whatever, under the designation "chocolate", that is not manufactured exclusively from shelled cacao, and that in a minimum proportion of 35%, and ordinary sugar, with or without admixture of spices.

Products which though containing the requisite 35% of shelled cacao are also made of other substances than those above signified may only be sold, held in possession, exposed or transmitted for sale under a label that clearly describes the nature of such ingredients next to the word "chocolate" and in the same type, or under a label that does not contain the word "Chocolate" at all. In the case indicated by impressing them on each separate tablet.

Products which contain less than 35% of cacao may only be sold, held in possession, exposed, or transmitted for sale under the designation "cacao bonbons" or some similar description, from which the word chocolate has been rigidly excluded.

Art. 3. Entries of the labels prescribed for the products of irregular composition in articles 1 and 2 must be made on the invoices despatched with the goods.

Art. 4. The box, case or wrapper etc. containing cacao or chocolate which is sold, exposed, held in possession or transmitted for sale must bear the name and address of the manufacturer or seller, or at least some regular and authorised trade mark.

Art. 5. The articles of this decree, as far as they refer to chocolate, are only applicable to ordinary chocolates in tablet, block, spherical or powder form, not however to cream and various sugar confections in chocolates (such as pralinés, pastilles etc.).

Art. 6. Any infringement of the articles of this decree will incur a fine in accordance with the code of fines issued on Aug. 4th 1890, over and above the ordinary penalties.

Art. 7. Our Board of Trade and Agriculture is hereby entrusted with the carrying out of this decree, which shall come into force on April 1st 1895.

2. Roumania.

The royal enactment of this land respecting the health supervision of foods and drinks and the trade in foods and drinks, articles 154, 155, 156 and 157 of the Health act of the 11th September, 1895, says the "Buletinul directiunei generale a serviciului sanitar" 1895, No. 18 and 19, pages 277 et seq.

No. XIII, Article 137.

No product may be sold, exposed or held in possession or transmitted for sale, under the designation cacao, other than the seed of the fruit obtained from the tree "Theobroma Cacao It may be brought on the market raw, roasted, or powdered after roasting

Under the designation "Cocoa powder, defatted", such may be sold as has suffered loss of butter by extraction, provided that there still remains a minimum 22% of cacao butter in the product. As disintegrated cacao may be sold such powder as does not contain more than a maximum 2% of sodium or potassium carbonate.

Art. 138. It is illegal to sell or expose for sale artificially dyed and pulverised cacao, and also such as has been mixed with starch meals, foreign fats or any other foreign ingredients. It is in like manner illegal to mix cocoa powder with shells, and the former may not contain more than a maximum 15% of powdered shell.[223]

Art. 139. Under the designation "Chocolate", only the foodstuff prepared from a mixture of roasted and powdered bean and sugar, with or without admixture of aromatic ingredients, as vanilla, cinnamon and the like substances, may be sold and exposed for sale.

Art. 140. The manufacture and sale, as also the exposure for sale of chocolate from cacao that does not answer the several demands of this decree, articles 137 and 138, as well as of chocolate that is mixed with starch, meals, mineral and artificially coloured substance, is illegal.

3. Switzerland.

The association of analytic chemists in this country have issued a book entitled "The Swiss Book of Nutritious Stuffs and Articles of Sustenance", where the methods and standards prevailing in research work connected with such substances are finally established for Switzerland. This work served as a guide as regards articles of sustenance up to the time when the Swiss food act came into force, and we accordingly annex a few extracts from it, dealing with our subject, cacao preparations.

Definitions.

1. Cacao mass is obtained by grinding and moulding the shelled and roasted cacao bean, without any admixture

whatever, or extraction of butter.

2. Cacao f r e e d   o f   o i l is cacao that has been reduced by from 20% to 35% as regards its butter contents by means of pressure under heat.

3. D i s i n t e g r a t e d   c a c a o. The roasted beans are treated with carbonic acid alkalis (generally potassium) subjected to pressure under ammonia or steam, and so the cellular tissue of the albuminous substance disintegrated or broken up and converted into a soluble modification (peptone and alkalinous albuminate).[224] The so treated beans are next dried, reduced, defatted and pulverised.

4. C h o c o l a t e is the description of a mixture of cacao and sugar which comes into commerce either moulded or in powder form. The percentage of sugar amounts to between 40 and 70%. Admixture of other substances than cacao, sugar and the usual spices must be regarded as adulterations.[225]

5. C h o c o l a t e and c a c a o (powdered or moulded) may be aromatised with the following substances: vanilla, benjamin gum, tolu and peru balsam, cinnamon, cloves and nutmeg.

6. C h o c o l a t e   f o n d a n t s are chocolates with an unusually large proportion of sugar and fatty contents.

7. M i l k   c h o c o l a t e is a preparation prepared from milk, sugar and cacao. It may not contain the preserving materials dis-allowed for milk, such as boracic acid, borax, formic aldehyde and derivatives of the aromatic series. It comes into commerce in powder form.

8. C o v e r i n g   or   c o a t i n g material is a mixture of cacao, sugar, spices, with almonds and hazel nuts. This preparation is almost exclusively employed for bonbon

confectionery.

9. Medicinal Chocolate is a chocolate or cacao preparation containing additions of medicaments.

Tests and Definitions always to be applied.

1. Touch test.
2. Reaction.
3. Microscopical examination.
4. Examination of the fat.
5. Estimation of cacao butter.
6. Determination of sugar.
7. Determination of ash.

Tests and Definitions eventually necessary.

8. Determination of moisture.
9. Determination of theobromine.
10. Determination of starch.
11. Determination of cellulose.

Guide to Classification:

Unripe, badly fermented cacao beans and those which have been attacked by insects or mould or have suffered during transport from the influence of salt-water, should never be used for manufacturing purposes.

Goods prepared from such beans have an unpleasant taste, which it is impossible to get rid of by the various operations in the course of manufacture. The use of all such beans is to be regarded as adulteration. The tests to be applied for determining them are tasting, microscopical examination and perhaps the estimation of the common salt contained in them.

All good chocolates are of a fine brown colour. Grey-coloured or spotted chocolate are objectionable. Spots or the grey colour alluded to may be caused either by damp or heat. At an ordinary temperature the fracture of the chocolate is hard, glassy and even. The quality of the fracture constitutes an excellent basis in judging of the manner and methods employed in working up the raw material.

Cacao and chocolate that become thick and pulpy on boiling are in all probability adulterated with meal, starch, dextrine or resin.

The following are to be considered as adulterations:

1. Admixtures of cacao or other shells, and sawdust.

2. Admixtures of foreign starch, meals, castania and resin.

3. Admixtures of mineral substances like ochre, clay and sand.

4. The substitution of cheaper fats, such as beef and pork dripping, almond, poppy seed, cocoa-nut and vaseline oils.

Limitations.

1. For cacao material

Ash	Maximum: 5% (Porto Cabello 4·65%)[226]
	Minimum: 2% (Surinam 2·25%)
Cacao butter	Maximum: 54·5% (Machalla 54·06)
	Minimum: 48·0% (Porto Cabello 45·87).[226]

2. For cacao fat. Melting point 29 to 33·5° C.; freezing point 24 to 25° C.; refraction at 40° C., 46 to 49[226]; iodine value 34 to 37; point of saponification, 192 to 202.

3. Disintegrated cacao the amount of added alkali is not to exceed 3%. In no case shall the ash content be more

than 8%. This figure is not inconsistent with the above stated maximum ash content, as disintegrated or soluble cacao is manufactured from a mixture of several sorts of cacao, in each of which (although they have been defatted) there is not more than 5% of ash.

4. C h o c o l a t e: although at the present time there are no limits fixed for cacao and sugar, it may nevertheless be safely assumed that the fat and sugar together may not exceed 80 to 85%, and that the rest shall be pure non-fatty cacao material, in the proportion of from 15-20%. The ash in a good chocolate does not exceed 3·5%.

5. M i l k c h o c o l a t e here the separate ingredients require a thorough drying. If the percentage of moisture amounts to as much as five percent, the whole preparation is objectionable and liable to lose its hard consistency.

6. C h o c o l a t e à la noisette, o a t, m e a t and m e d i c i n a l c h o c o l a t e s The testing of these takes two chief directions:

1. It must be established that the ingredients given on the label are of good quality, and

2. that only the ingredients there mentioned occur in the packet.

The constituents and their proportions shall be declared on the wrappers in the case of medicinal chocolate.

On the 1st July then, in the year 1909, the act passed in connection with foods and articles of consumption December 5th, 1905 came into force in Switzerland. Thereby the whole of Swiss trade in such foodstuffs and articles of consumption is systematically controlled. Of the 268 articles which are generally representative, we annex here those concerning cacao, powder and chocolate, namely, nos 146 to

149.

Art. 146. Under the designation c a c a o or c a c a o p o w d e r only the pure, unaltered or only partially defatted natural product may be brought into commerce.

A cacao powder may only be described as s o l u b l e when it has been treated with carbonic acid alkalis or disintegrated with steam.

Soluble cacao may only contain 3% added alkalis on the outside.

Art. 147. Under the designation c h o c o l a t e, only a mixture of cacao and sugar with or without addition of cacao butter and spices is to be understood, and no other may be brought on the market as such.

The percentage of sugar in chocolate may not exceed 68.[227]

Art. 148. Cacao and chocolate may not contain starch, meal, foreign fat, mineral substances, colouring matter and so-called fat economisers (dextrine, gelatine, resin and tragacanth) and only traces of cacao shell. They may not be gritty nor foul smelling nor otherwise spoilt.

Art. 149. Special products of cacao and chocolate with addition of oats, milk, acorns and hazel nuts must be declared accordingly (as oat cacao, milk chocolate etc.). Fancy confections fall also under this obligation.

Cacaos and chocolates which are put on the market in packets, boxes and packages must contain the name of the firm on the wrapper, or some mark of the manufacturer or salesman which is recognised in Switzerland.

If saccharine, dulcine or other artificial sweetstuffs are added to chocolate, such admixture must be declared on the wrapper.[228]

4. Austria.

Legal control of the traffic in cacao preparations in this country may be expected in the near future.

A u s t r i a is indeed already in possession of a law (dated January 19th, 1896) concerning the traffic in articles of consumption, although the special determinations have hitherto not reached perfection, and the treatment of the separate detailed articles must proceed gradually. As in Switzerland, the Association of Food Chemists and Analysts here have worked out designs for a "Codex alimentarius austriacus The work of this code commission is of a purely private nature and accordingly no official importance accrues to it, but it is none the less recognised by all Austrian chemists and has indirectly (and even in law courts) about the same weight as the opinion of an expert, especially as the single articles of consumption are almost exclusively limited to specialists in this country. We therefore introduce the most important points of this code which bear on our subject, although various alterations must be made in these as they succeed to legal recognition, for since the appearance of the code many changes have developed as regards the methods of research.

I. Cacao Mass.

D e fi n i t i o n . Under cacao mass is to be understood the material constituting a regular and uniform dough when warmed, which has been exclusively prepared and manufactured from the shelled cacao bean.

I n g r e d i e n t s . Cacao material contains the same

ingredients chemically as the shelled bean.

Microscopical investigation should only reveal the presence of seed kernel, and not particles of root, which should be removed in the course of preparation.

The ash may not exceed 3·5%[229], the fibre 3%[229], and the starch 10·5%. The amount of fat figures at between 48 and 52 percent.

II. Cocoa powder.

(Pulverised cacao, defatted, and disintegrated.)

D e f i n i t i o n . Hereby is understood the steamed preparations or the powder obtained by expressing at least half the total fat from ordinary cacao material and further grinding and sifting.

C h a r a c t e r i s t i c s . The cocoa powder shall on boiling with 20 to 30 times its volume of water yield a suspension, in which there are no traces of lumpy formation, and which does not show a sediment after the expiration of a few minutes.[230] Should there be any such sediment, it shall be examined under the microscope.

Cocoa powder shall be sifted and ground free from meal, and may not, on sifting through a miller gauze (No. 12) show more than 5% of material on the sieve.

The chemical composition of cocoa powder is modified according to the degree of defatting. If 30 parts out of 100 are defatted, which is the usual procedure. If 30 parts fat are expressed from 100 parts cacao material, which usually happens, then the cocoa powder contains

30% fat, 5% ash[231], 3·5% fibre, and 13%.

The amount of moisture shall not exceed 6%.

The fat shall be pure cacao butter.

Addition of alkalis is not allowed.

Microscopical investigation as under I.

III. Chocolate.

D e f i n i t i o n . Chocolate is the cacao material evenly and regularly worked up with cane sugar (refined, ordinary or coarse).

The completely uniform pasty mass, when warmed, is allowed to set in moulds and then forms pieces of fatty appearance, finely granular or close fracture (tablets, blocks).

Good chocolate consists of 40 to 50 percent of cacao mass and 50 to 60 percent of sugar.

It may also contain a small amount of harmless aromatic substances.

Should the sum of the cacao fat and sugar in chocolate amount to over 85 percent, it is termed "Sweetmeat chocolate", and should the sum of those ingredients be more than 90 percent, the chocolate is to be declared as "Very sweet

All the ingredients in chocolate, after deducting the sugar, shall be present in the same relative proportion and in the same condition as in pure cacao mass (compare I).

Sweetened chocolate is an exception, in so far as it has had in its preparation an addition of cacao butter. Fine kinds are also prepared with an addition of defatted cacao.

U n m o u l d e d chocolate or chocolate powder shall answer to the same requirements.

416

IV. Cacao surrogate and chocolate surrogate.

D e f i n i t i o n . Cacao preparations containing admixtures of meal are to be described as surrogates.

The addition of other substances than meal is inadmissible.

Absence of cacao husk is also required as in I, II, III.

Mixtures of cacao powder, sugar and meal are also to be regarded as surrogates.

The extent of the addition of meal is to be distinctly noted by the seller on the article sold.

V. Couverture (coating mass).

D e f i n i t i o n . This includes various preparations of pure cacao butter and chocolate (or mixtures of chocolate with cacao butter and cacao mass), which form a thin liquid, when warmed, and are used for coating or pouring over confectionery. All other substances (roasted hazel nuts or almonds and the like) shall be declared.

Investigation.

T o   b e   c a r r i e d   o u t   w i t h o u t   e x c e p t i o n w i t h   a l l   c a c a o   p r e p a r a t i o n s

1. D e t e r m i n a t i o n   o f   f a t . The fat is extracted from the dry substance which has been mixed with an indifferent body (sand) by pure and absolutely dry ether (distilled over sodium) or by petroleum ether. Cacao mass and chocolate must first be shaved or rasped.

2. J e s t i n g   o f   t h e   f a t .

a) Determination of the melting point in a capillary tube (three days after the fat has been melted into the tube).[232]

Pure cacao butter usually melts at 33° C.

b) Determination of the iodine value; usually 35·0 with pure cacao fat.

It is further recommended to make a refractometric determination, which in a Zeiss butter-refractometer must be 46·5° at 40° C.

3. The microscopic test of the substance, from which the fat and the sugar have been removed.

The following are also essential

I. With cacao mass.

The determination of fibre and ash.

II. Cacao powder.

Determination of moisture at 100° C., of the fibre and ash and examination of the ash (quantitative determination of phosphoric acid and potash).

III. Chocolate.

Determination of the sugar by polarisation of the aqueous solution.

IV. Surrogates.

Determination of the starch.

If it is considered necessary to proceed further, then:

1. Determination of theobromine by a modification of Wolfram's method, the method employed is to be exactly stated.[233]

2. In the determination of starch, the gelatinisation is to be carried out under steam pressure and the inverted sugar

gravimetrically determined with Fehling's solution.

An opinion of the quality of the preparation can be formed from the taste, smell and colour of the sample on boiling with water.

5. Germany.

In G e r m a n y, unfortunately, there is at present no law, which regulates the trade in cacao goods. It is true that there exists the decree of the 14th May, 1879 respecting the trade in food, alimentary substances and comestibles, which contains the usual penal enactments in regard to adulteration of food materials offered for sale. The enactments are supplemented with data relating to the administration of the law, among which a definition of chocolate, as well as the means of judging as to the quality or its adulteration, are treated of. But those data do not in all respects apply to existing conditions, nor do they deal fully with the question as to what admixtures are to be permitted or prohibited, for in the introduction to the appendix A, there is the following statement:

"Like the former provision, the present one is not intended to be an e x h a u s t i v e description of all subjects of the kind referred to, but a compilation of those examples which appear to be especially calculated to serve as an illustration of legislative requirements."

The data referred to have not an officially authoritative significance, and they cannot be regarded as having established validity in connection with the administration of the law by the police or by legal authorities. (See: Commentary by Meyer-Finkenburg, page 116.)

Even the complete publication of the "Vereinbarungen zur einheitlichen Untersuchung von Nahrungs-und Genußmitteln sowie Gebrauchsgegenständen für das Deutsche Reich", collected at the instance of the national health department, will not have the effect of giving certainty in the law relating to the manufacture of chocolate. That section of the "Vereinbarungen", which deals with cacao products, was published in Book III (Berlin, Julius Springer 1912) pages 68-81, but the conditions in Germany are at present only similar to those existing in Switzerland and in Austria. The "Vereinbarungen" are nothing more, than a valuable semiofficial guide for the valuation and examination of food and comestibles, the provisions of which, not being obligationary, have no legal effect. They have long been in need of a thorough revision, as recent scientific results testify, and indeed "The Voluntary association of German Food Chemists" have for years been engaged in such revision.

The consequence is, that the prosecution of various manipulation which certainly deserve to be objected to, such as the preparation of cacao or chocolate from undecorticated beans, would be difficult to carry out. The Association of German Chocolate Manufacturers has protested against that unsuitable state of affairs, and since a remedy is to be looked for only from the enactment of a law regulating the trade in cacao products, that association prepared a draft act, at its XVII. annual meeting at Leipsic on the 15th January 1893, and has submitted it to the government health department.

That draft is in accordance with the provisions printed on pages 231 and 232 a-e. The provisions of the association in reference to the trade in cacao products also contain the following paragraphs:

§ 2.

It is not to be considered adulteration or counterfeit, within the meaning of the law (§ 10) relating to trade in food materials, comestibles or articles of consumption (of 14th May 1879, Reichsgesetzblatt page 145):

1. When the productions referred to under a, b, c are mixed with meal or other substances for medicinal purposes, provided, they are of a character by which they are distinctly recognisable, or are kept in stock or offered for sale under a designation distinguishing them from chocolate, cacao mass, or cacao powder.

2. When covering or coating material, or sweetmeat chocolate is mixed with burnt almonds or hazel nuts to the extent of 5 %.

§ 3.

Adulteration within the meaning of the law dated May 14th 1879, § 10 (Reichsgesetzblatt, page 145) comprises:

1. The addition of foreign fat to chocolate, cacao mass or cacao butter.

2. The addition to chocolate, cacao material or cocoa powder of cacao husk, meal or other substances, except in the cases mentioned on page 279, § 2, pos. 1 and 2.[234]

3. The addition of colouring materials to chocolate.

4. The addition to chocolate or chocolate surrogates of any but cane sugars (beetroot sugar).

§ 4.

As already pointed out, the terms of this proposed legislative step naturally command approval and we should be the first to welcome the appearance of a "Deutsches

Lebensmittelbuch" or some similar work[235], intended to serve as an authoritative regulation of the trade in cacao preparations and as a protection of honest manufacturers against the uncertainty now attending legal proceedings. In that case, other civilised countries might be expected to follow.

Book 5.
Appendix.

A. Installation of a chocolate and cacao powder factory.

In constructing a new factory and fixing the situation of the buildings, the first thing to be considered is their convenient arrangement. It is therefore advisable to rely upon an experienced person for the plan to be adopted, and then to leave the proper construction of the works in the hands of the architect. Small operations can be carried on in any building, but in the case of larger works a well devised arrangement of the machines and appliances must be decided upon before hand, that will admit of rational and, to some extent, automatic working. In case of erecting small works which will require only one manager, the best plan would be to have the whole manufacture carried out on one story, or at the most two stories, to facilitate supervision.

The case is different with large works, in which the different departments are controlled by especially qualified persons.

Tables I and II[236] represent, in section, a chocolate factory and a cacao powder factory. As both plans represent only a model section, they serve only to show the most convenient arrangement of the machines with each other. In reality there would be more or less machines of the same kind placed together. Such arrangements might, with modifications, serve for medium sized works, as well as for larger ones. In that sense the following explanations of the two plans are to be understood.

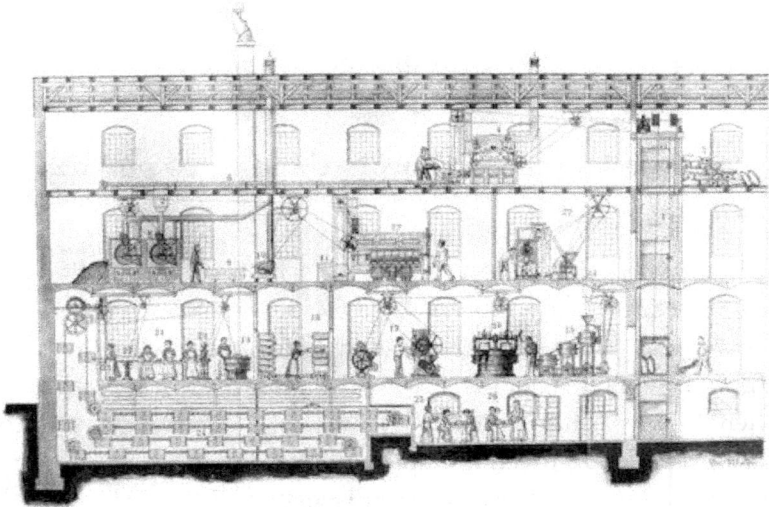

PLATE II

Longitudinal Section of a model Chocolate Factory
For explanation of figures see text.

Zipperer, Manufacture of Chocolate etc. 3rd edition.
Verlag M. Krayn, Berlin W. 10.

Click on the images to see a larger version.

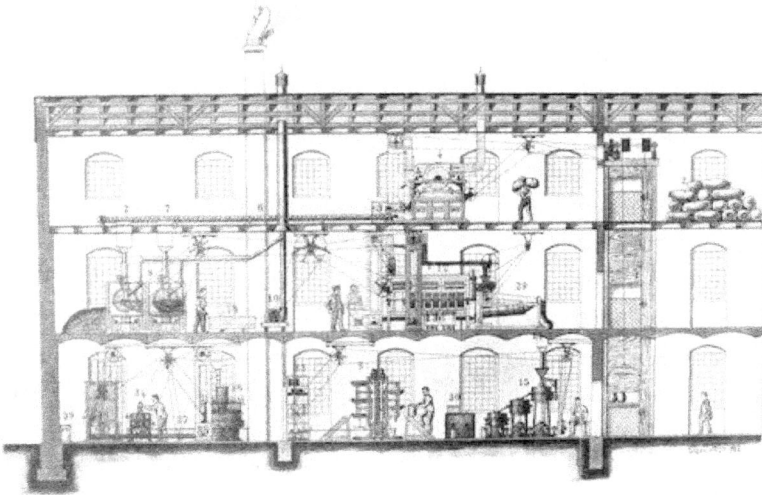

PLATE III

Longitudinal Section of a model Cocoa Factory
For explanation of figures see text.

Zipperer, Manufacture of Chocolate etc. 3rd edition.
Verlag M. Krayn, Berlin W. 10.

1. Chocolate factory (Table I).

By means of the lift (1) all the raw materials, sugar, cacao, packing materials, etc. are carried up to the store rooms (2). In these occur the machines for cleansing and picking the raw cacao beans. The raw cacao is fed into the elevator boxes (3), above the cleansing machine (4) where it is freed from dust; it passes to the continuous band (5) where it is picked and then falls into the movable boxes (6). It is then transferred to the hoppers (7) and is fed, by opening a slide in the hoppers, into the roasting machine (8). The capacity of the hoppers is sufficiently large for holding the quantity of beans for charging the roasting machine. After the roasting is completed, the cacao is emptied into the trucks (9) and carried to the exhaust arrangement (10) where the

425

beans are cooled down and the vapour given off is passed out into the open air. At the same time, the roasting chamber is sucked out through the funnel shaped tube fitted to the cover of the chamber. The roasted cacao is then passed to the boxes (11) to be conveyed by the elevator to the crushing and cleansing machine (12). After being cleansed, the cacao is carried in trucks (13) to the hoppers (14) by which they are fed into the mills (15) in the lower floor. The sugar mill and the sifting apparatus (27) placed near the crushing and cleansing machines are also fed by a hopper from above. The dust sugar, there produced, is carried by the lift (1) to the machine room on the first floor. Cacao and sugar are there supplied to the incorporator (16) to be worked together, before being passed to the rolling mill (17), where the final rubbing is effected. After passing once or oftener through the mill, the finished chocolate mass is then taken to the hot room (18) where it remains in boxes until further treated and it is then taken to the moulding room. In the incorporator (19) the mass acquires the consistence necessary for moulding and also the requisite temperature. The mass is then taken in lumps to the dividing machine (20) and cut into pieces of the desired size and weight. On the table (21) the moulds, lying upon boards, are filled with the pieces of chocolate and they are then taken to the shaking table (22).

From this they succeed to the cooling arrangement, which consists of an endless chain provided with travelling stages at definite and regular intervals. The latter moves slowly through the artificially cooled room and finally brings the moulds to the outlet (25) where the chocolate is removed. It is then transferred on the lift to the packing and despatching apartments specially reserved for these operations, but not distinctly noticeable on our section.

2. Cacao powder factory (Table II).

The course of manufacture of cacao powder is the same as in the manufacture of chocolate, up to the point where the cacao has passed through the crushing and cleansing machines (12). The broken beans are then taken by the elevator (27) to the machine for separating the radicles (28) and thence through the hopper (14) to the mills (15). The liquid cacao mass, passing from these mills, runs into the pans (29) from which as much required for charging the hydraulic presses as is can be drawn up by cocks. The accumulator (31) supplies all the presses with water. The pressed cakes are first put into the boxes of the frame (32). In an adjoining room is the automatic cacao pulverizing apparatus. It is fed through the preliminary crusher (34) from which the cacao is taken by the worm and elevator (35) to the pulveriser (36). The powdered cacao is then taken by a worm and elevator to the sifting machine (38).

The sifted powder falls into the tub (39) while the coarser portion is carried back again to the pulveriser (36). The arrangements for treating and the disintegrating cacao powder can be provided in the manner already described.

In both plans, the boiler and engine house are to be understood as placed in an adjoining building.

Appendix

Containing an account of the methods of preparation and the composition of some Commercial dietetic and other Cacao preparations.

The following statements and recipes have no pretension

to be complete; they are only introduced to serve as a brief summary of those commercial cacao preparations, now in commerce, which are mixtures of various kinds of substances with cacao or chocolate and are largely used for dietetic purposes. Notwithstanding its necessary incompleteness, the following account, which has been collected from various sources, will satisfy practical requirement, since the manufacturer, as well as the food chemist, frequently desires to obtain information at once, that even a complete technical library is not always able to supply. Medicinal chocolates have not been considered in the following list, since they belong to the province of pharmacy.

>

Acorn-cacao Michaelis' contains according to an analysis by R. Fresenius: Total nitrogen 2·29 percent, albumin 8·13 percent, sugar 25·17 percent, starch 23·39 percent, fat 14·42 percent, tannin, expressed as gallotannic acid 1·96 percent, cellulose 1·88 percent.

Acorn-cacao of Hartwig & Vogel of Dresden contains water 7·5 percent, ash 3·88 percent, fat 16·54 percent, albumin 11·25 percent, carbohydrates 38·76 percent, tannin 2·50 percent.[237]

Acorn-cacao of Th. Timpe of Magdeburg contains in the dry substance: albumin 13·88 percent, tannin and cacao-red 5·37 percent, carbohydrates etc. 66·41 percent, fat 10·62 percent, ash 3·73 percent.[238]

Acorn-cacao can be prepared by mixing 10 parts of pure cacao mass, 20 parts defatted cacao powder, 5 parts roasted barley meal, 35 parts of the meal from shelled and roasted acorns (or 10 parts of an aqueous extract of roasted acorns), 30 parts powdered sugar, and 2 parts pure calcium phosphate.

Acorn-chocolate is a mixture of 100 parts shelled and roasted acorns with 500 parts sugar and 400 parts cacao mass in addition to spices.

Acorn-malt-cacao (Dieterich) is prepared by mixing 1 kilo of acorn malt extract (Dieterich-Helfenberg) with 6 kilos of sugar (dust), and 3 kilos defatted cacao.

Acorn-malt-chocolate (Dieterich) is prepared by accurately mixing 2 kilos acorn malt extract (Dieterich-Helfenberg) with 3½ kilos of powdered sugar and 4½ kilos of cacao mass.

Albuminous chocolate and cacao. Riquet & Co. of Leipsic have protected a process by various patents[239] for "The production of a tasty and genuine chocolate or cocoa powder[240] rich in albuminous constituents." The kernels of the thoroughly roasted bean are worked up with a mixture (?) of water and dry albumen, allowed to stand for some time, the water evaporating, and then the beans are worked up once more. Instead of water an aqueous sugar solution may also be employed, and further the addition of albumen may occur at any stage[241] and in particular when sugar solution is first taken, then the albumen and sugar necessary for the chocolate mixed up, and finally the cacao material (with additions of cacao oil) added. Still better (than the sugar solution) would it be, if the albumen were incorporated in the chocolate or cocoa material in the form of a mixture with some emulsion (!), especially a mixture with milk.

Barley-chocolate is prepared by mixing 1 kilo of prepared barley meal[242] 4½ kilos powdered sugar and 4½ kilos cacao mass. The moulded chocolate is to be coated with varnish.

429

Cacao and chocolate preparations containing milk are prepared according to A. Denayer, Brussels (German patent No. 112220, 4 February 1899) by evaporating, in the open air, a mixture of milk and sugar to the consistency of cream, and to the hot mass, defatted or not defatted cacao is added in the form of powder. The resulting mixture is spread out in thin layers and exposed to the influence of a temperature of 80-100°C. in a rarefied atmosphere, then finally completely dried at a lower or ordinary temperature under the same conditions.

Cacao-egg-cream (so called African punch) is thus prepared: 10 yolks of eggs are beaten up with 300 grammes of syrup (1 part sugar to 2 parts water) and, whilst being continually whisked up, 500 grammes of cacao essence (see next paragraph) are added. The whole is to be iced before being consumed.

Cacao-essence is prepared by macerating 125 grammes of defatted cacao, 2 grammes vanilla, 2 grammes cinnamon, 0·75 gramme cloves, 0·3 gramme mace and 0·10 gramme of ginger with 750 grammes of proof spirit and 250 grammes of water for 8 days, and then filtering into hot syrup, which is prepared with 550 grammes of sugar and 750 grammes water.

Cacao-liqueur. A well tested recipe for the preparation of this liqueur is to the following effect: Defatted cacao 200 grammes, cinnamon powder 5 grammes, vanillin 0·2 gramme, are digested for 6 days with 1500 grammes of water and 1700 grammes of alcohol (90%) and then mixed with 2600 grammes syrup (1400 parts sugar and 1200 parts of water) and filtered.

Cacaol, 70 parts cocoa powder, 10 parts oatmeal, 17·5 parts sugar, 2·5 parts common salt.

Cacao-malt is a mixture of 20ℑ parts defatted cacao, 500 parts sugar with an aqueous extract of 300 parts of kiln dried malt.

Cacaophen Sieberts (Cassel) is a mixture of cacao powder with flour, sugar and milk albumin. It shows the following numbers on analysis: fat 13·23 percent, water 7·7 percent, albumin 24·25 percent, soluble carbohydrates 17·95 percent, insoluble carbohydrates (starch) 26·66 percent, woody fibre 2·27 percent, ash 5·5 percent (calcium oxide 0·82 percent, phosphoric acid (P_3O_5) 0·54 percent).

Children's-Nährpulver (Lehmann-Berlin) is a mixture of meat extract, cacao powder, salep, sugar and specially treated oyster shells.

Chocleau, (Reichardt) a glucose chocolate material in tin tubes.

Chocolate-cream-syrup (for aërated waters): 125 grammes of rasped chocolate, 62 grammes cacao powder and 325 grammes of water are well mixed and to this add 148 grammes infusion of quillaia (1·8). After standing some time add the contents of a pot of condensed milk with 7·5 grammes of boric acid and make up with 3·8 litres of sugar syrup (american recipe).

Chocolat digestif (Vichy chocolate) is a mixture of chocolate with about 5 percent of sodium bicarbonate.

Chocolate-health-beer, J. Scholz (German patent No. 28819). An extract is prepared from 10 kilos of cacao beans, which have been kiln-dried at 75° C., shelled, broken in small pieces and digested for half an hour with twice their weight of distilled water at 62° C., then boiled for another half an hour and finally allowed to stand for 48 hours at a temperature of 75° C., with an addition of a

431

solution of 10 kilos of sugar in distilled water, then once more boiled until one half of the water, originally added, has been evaporated. It is filtered, in as warm a condition as possible, in order to separate pieces of cacao and fat, and the extract is ready for use. The brewing process is similar to that of brewing Bavarian beer. After the finished wort obtained in that process has been boiled for 3 hours, 100 litres are taken, for which 35 kilos of pale kiln-dried barley meal have been used, and to this are added 200 grammes of the best Bavarian hops and 12 kilos of cacao extract. The whole is once more boiled and the subsequent operation then carried out as usual. The fermentation (at 7·5° C.) occupies 7-8 days and the storage in the fining vats 3-4 weeks.

Chocolat rétablière, a Vienna speciality, contains reduced metallic iron, dried meat, pea and wheat flour, sugar and cacao in uncertain proportions.

Chocolate-syrup (for soda and seltzer water). 250 grammes of defatted cacao powder are rubbed down with 2½ litres of boiling water in a porcelain basin on a steam bath, until it is in the condition of an uniformly thick mass and then 1 kilo pot of condensed milk and 2·5 kilos of powdered sugar are added, and when the sugar is dissolved the vessel is cooled. After cooling, the fatty particles on the surface are carefully removed, and then 30 grammes of commercial vanilla extract and 30 grammes of mucilage (from gummi arabicum) are added, and the whole filtered through a stout cotton cloth (american recipe).

Chocolate-tincture (cacao-tincture) is prepared by macerating 1½ kilos of defatted cacao powder with 10 kilos of dilute alcohol for 8 days and then filtering.

Corn-cacao contains according to Notnagel[243]: water 6·10 percent, fat 16·96 percent, albuminoids 19·81 percent, theobromine 0·68 percent, fibre 3·30 percent, non-nitrogenous extractives 48·69 percent, ash 4·46 percent. The preparation under the microscope is shown to contain, in addition to the constituents of cacao, a large amount of oat starch, and it may be regarded as corresponding to a mixture of equal parts of defatted cacao and oat meal, based on the above analysis and König's mean value.

Covering or coating materials have the following composition: 50% sugar, 30-35% fat and 20-15% cacao material free from fat, whereby (especially in Belgium, e. g. Brussels) it is in part supplanted by almonds, nuts etc. In such cases the iodine value of the fat is equal to 41-42.

Diabetic chocolate has the following composition.[244] Nitrogenous substance 10·07 percent, fat 25·47 percent, levulose 19·38 per cent, starch and cellulose 25·19 percent, besides non nitrogenous substances 14·54 percent, saccharin 0·5 percent, mineral constituents 2·15 percent.

In this formula there is a disproportionately high percentage of starch and cellulose and, in that respect, the composition appears to be irrational, since the introduction of carbohydrates into food for diabetics should be avoided as much as possible. A more rational preparation would be a simple mixture of:

50 parts levulose | 50 parts cacao mass,
and 0·25 parts vanillin.

Aufrecht's recipe for **diabetic cacao** is as follows:

cocoa powder	500 grammes
levulose	200 "
wheat flour	280 "

saccharin   5   "

aromatic substances   15   "

In this recipe, also, the substitution of levulose for wheat meal is to be recommended.

Diabetic cacao can be prepared according to J. Apt of Berlin by the following patented process (German patent No. 116 173, 30. 1. 1900). The starch is first gelatinised by long boiling of the coarsely powdered cacao, the mass then dried in a vacuum and heated, or roasted at 130 to 140° C. in order to caramelise the gelatinised starch (!). Before being boiled, it is recommended to de-fat the cacao (with petroleum ether, for example!). Instead of caramelising the gelatinised starch by heat direct, it can be first converted into sugar by means of acid, then heated to caramelisation and as much cacao fat added as may be desirable. In order to increase its capability of emulsifying, dried albumin is to be added.[245]

Dictamnia of Groult and Boutron-Russel is composed of cacao, prepared wheat flour, starch, sugar and vanilla.

v. Donat's albumin chocolate (German patent No. 82 434) is prepared by mixing dried albumin in powder or in pieces with chocolate or cacao mass, damped with a liquid medium, which does not dissolve albumin, such as benzol, petroleum ether, ether, acetone, methyl or ethyl alcohol. The mass is further treated in the mixer and finally after being completely mixed, the added liquid is allowed to evaporate.

Eucasin-chocolate and cacao are preparations containing 20 percent of eucasin (ammonium caseinate). Eucasin is prepared by Majert & Ebers of Grünau-Berlin.

Galactogen-Cacao, Thiele & Holzhause-Barleben near

Magdeburg, contains 30-32 percent of galactogen, an easily soluble and natural preparation of milk albumin, which is prepared from skimmed milk and contains 70 percent albumin, 3·5-4 percent fat as well as 1·5-1·79 percent phosphoric acid. G a l a c t o g e n - a m y l a c e o u s c a c a o, contains wheaten flour in addition to 20-22 percent galactogen. Galactogen-Speise-Schokolade (eating chocolate with 30 percent galactogen and Galactogen-Koch-Schokolade (cooking chocolate) are also prepared.

Plasmon, Jropon, Somatose and lacto-egg-powder are similar products to galactogen, and are met with in commerce combined with cacao mass and chocolate (see plasmon cacao).

Gaugau is a children's tea (Vienna) and consists of cacao husk.

Haema chocolate: 25-30 parts cocoa powder, 25-20 parts meal (potato starch), 45 parts sugar, 5 parts haemoglobin and common salt.

Hansa-Saccharin-Cacao is defatted cacao, which contains about 0·5 percent saccharin (270 times as sweet as sugar), 30 percent fat and 20 percent albuminoids (Hahn-Holfert).

Hardidalik, an Asiatic chocolate, is composed according to Chevallier of 42 parts cacao, 180 parts sugar, 112 parts starch flour, 64 parts rice flour and 3 parts vanilla.

Hensel's Nähr-Cacao, is a mixture of defatted cacao-powder with various inorganic salts, such as calcium carbonate and phosphate; the ash of this preparation was found to contain a larger amount of sulphuric acid, soda and iron, than is present in normal cacao. The fat amounted to only 5·3 percent.

Homeopathic-Chocolate of E. Kreplin, Lehrte, consists of 35 percent pure cacao mass, 20 percent slightly roasted wheat flour and 45 percent sugar (Hager).

Husson's Mixture contains the following materials: Arrow root 500, oat meal 500, powdered sugar 500, powdered sago 400, cacao 50, calcium phosphate 50, vanilla 1.

Hygiama resembles cacao in appearance and flavour and was introduced into commerce by Dr. Theinhardt's Nahrungsmittel-Gesellschaft of Cannstatt (Wurtemberg). It is prepared from condensed milk with the addition of a specially prepared cereal and defatted cacao. It contains 22·8 percent of albumin, 6·6 per cent. fat, 52·8 percent soluble carbohydrates, 10·5 percent insoluble carbohydrates, 2·5 percent food salts, 4 percent moisture.

Iceland-moss-chocolate contains 10 percent of iceland moss gelatine.

Kaïffa (Fécule orientale) is a mixture of 500 parts cacao mass, 1250 parts rice flour, 250 parts groats, 250 parts Iceland moss gelatine, 2300 parts starch, 750 parts salep, 1000 parts sago, 6000 parts sugar and 50 parts vanilla.

Kola-Chocolate is prepared by mixing 400 grammes of cacao mass, 450 grammes sugar, 100 grammes kola seeds in powder, 40 grammes cacao fat and 5 grammes vanillin sugar (3 percent).

Kraft-Chocolate (Mering's). This is a trade preparation in which cacao butter is converted into an emulsion, probably by means of oleic acid, and is thus rendered more digestible. Kraft-chocolate should contain 21 percent of easily digestible fat.

Lipanin-Chocolate contains 42·38 percent fat, albumin 8·07 percent, starch 2·7 percent, sugar 31·44 percent, in addition to non-nitrogenous substances 18·19 percent, ash 0·68 percent, as well as some vanillin and Peruvian balsam (Aufrecht).

Malt-cacao according to Franz Abels (German patent No. 96 318, 9. May 1896) is prepared in the following manner: The cacao mass after being mixed with malt meal is defatted by strong hydraulic pressure in order that the malt may be permeated with cacao fat. It is then pulverized.

Malt-cacao-syrup or **malted chocolate** is prepared by mixing 240 grammes malt extract and 24 ccm vanilla extract with about 950 grammes of chocolate syrup. Vanillin or essence of cinnamon may be used instead of vanilla extract. This preparation serves for the making of american effervescing lemonade.

Malt-chocolate. 2 kilos of finely powdered malt and 3½ kilos powdered sugar, both well dried, are mixed in small quantities with 4½ kilos cacao mass in the mixing machine. The tablets are to be coated with varnish to preserve them. (E. Dieterich.)

Malt-extract-chocolate. 4½ kilos of the finely rubbed down cacao mass, contained in the mixing machine, are intimately mixed with 1 kilo dried malt extract and 4½ kilos powdered sugar. The finished tablets are to be coated with varnish. (E. Dieterich.)

Malto-leguminose-cacao gives the following numbers on analysis: water 7·38 percent, nitrogenous substance 19·71 percent (18·26 percent digestible), theobromine 0·71 percent, maltose 1·88 per cent., dextrin etc. 3·53 percent, starch 27·82 percent, besides non-nitrogenous extractives 13·8 percent, fibre 2·36 percent, ash 4·94 percent potash 1·74 percent, phosphoric acid 1·51 percent.

Meat-extract-chocolate is prepared by placing 500 grammes of meat extract (Cibil's or Liebig's) in a porcelain basin and evaporating as much as possible on the water bath: 4·7 kilos of powdered sugar are then added and the whole rubbed down with the pestle until the extract is homogeneous. 5 kilos of cacao mass are added and the chocolate finished in the mixer. The moulded tablets must be coated with varnish (Dieterich).

Milk-cacao is prepared with 1 kilo of condensed milk (prepared in a vacuum with the addition of 10 percent of milk-sugar[246] 500 grammes milk sugar and sufficient powdered arrowroot to produce a paste, which is then rolled out, broken up and lightly baked. This milk biscuit is ground and passed through a fine hair sieve. 750 grammes of the pulverized milk biscuit are then carefully mixed with 250 grammes of defatted cacao and 10 grammes of an aromatic mixture and the preparation finally preserved in metallic boxes.

A more bitter milk-cacao can also be prepared with 50 kilos cacao powder and 50 kilos pure milk powder. This proportion may also be varied, so that more milk powder may be used, as for example 40 kilos cacao powder and 60 kilos pure milk powder or 30 kilos cacao powder and 70 kilos pure milk powder.

A sweet-milk-cacao can be obtained thus:

a) 30 kilos		cacao powder,
20	"	powdered sugar,
50	"	pure milk powder
b) 20 kilos		cacao powder,
30	"	powdered sugar,
50	"	pure milk powder,
c) 15 kilos		cacao powder,

438

35 " powdered sugar,

50 " pure milk powder.

Milk-chocolate is prepared with 28 kilos of cacao mass, 36 kilos of powdered cane sugar, 24 kilos of milk powder and 12 kilos of cacao butter. The material is very finely rolled at 60-70°C. in the grinding machine described on page 000, and the finished mass not allowed to remain in the hot closet, but almost immediately moulded and packed. The mild kinds of cacao (Ariba, Caracas, Ceylon, Java) are the most suitable for making milk chocolate.

In the manufacture of p u r e m i l k c a c a o, the cacao powder is worked up for some time in the warmed mixing machine, the sugar and the milk powder being added successively. Cacao preparations, which are only used as beverages with water, should have at least two parts of pure milk powder to one part of cacao powder in order to yield a suitable preparation.

Mutase-cacao with 20 percent mutase: contains water 5·66 percent, fat 25·24 percent[247], albumin 28·31 percent, fibre 3·81 percent, theobromine 1·67 percent, non-nitrogenous extractives 30·72 per cent., ash 6·26 percent.

Mutase-chocolate (with 20 percent mutase) contains 16-17 percent of albumin. Mutase is an albumin preparation obtained, without the use of chemical reagents, from nutritive plants, also containing the nutritive salts of the plant (10 percent). Mutase contains 60 percent of albumin.

Nährsalz-cacao (Lahmann), i. e. "Food-salt cacao It contains water 8 percent, nitrogenous substance 17·5 percent, theobromine 1·78 percent, fat 28·26 percent, starch 11·09 percent, non-nitrogenous extractives 26·24 percent, fibre 4·21 percent, ash 4·7 percent (potash 1·66 percent, phosphoric acid 1·56 percent). N ä h r s a l z - c a c a o o r chocolate is prepared by mixing a vegetable extract (from leguminous plants) with cacao or chocolate. The analysis of L a h m a n n's N ä h r s a l z - c h o c o l a t e gave the following numbers: fat 24·5 percent, ash 1·36 percent, water 1·08 percent, albumin 6·25 percent, phosphoric acid (P_2O_5) 0·44 percent.

Nähr-und Heilpulver. (Food and health-powder) of **Dr. Koeben** contains sugar, cacao, pollards and acorn coffee. (Hager's Handbuch der Pharmaceutischen Praxis.).

Natur-cocoa and natur-chocolate (natural cacao etc.) Spindler, Stuttgart (German patent No. 47226) are obtained by mixing cacao mass with hot honey. This effects a defatting of the cacao mass by spontaneous separation of the fat. The defatting can be suitably carried further by pressing. Instead of using honey, the defatting can be carried out with syrups, malt extract, condensed milk, fruit juices or plant mucilage (extracts from pulse).[248]

Nuco-cocoa is a mixture of cacao with "nuco", which is a highly praised preparation of albumin. The analysis of nuco cacao gave ash 4·06 percent, moisture 6 percent, fat 15·23 percent, albumin 47 percent, the iodine value of the fat is = 86. The fragments of tissue under the microscope appear completely analogous to that of earth nut (arachis hypogaea). Nuco-cacao is consequently nothing more than a mixture of defatted cacao with defatted earth nut (earth nut cake).

Oat-cocoa, Hallenser (half and half) contains 6·5 percent moisture, 4·1 percent mineral constituents, 89·4 percent organic substances (containing 4·3 percent nitrogenous matter) digestible albumin 14·7 percent, fat 17·2 percent, theobromine 0·77 percent, starch and other non-nitrogenous extractives 48·93 percent, cellulose 3·5 percent. This is evidently a mixture of equal parts of oat meal and cacao powder as the name implies.

Oat-cacao Kasseler (Hansen & Co.) is prepared according to the German patent No. 93500, 28th June 1896, by mixing oat meal with cacao. This mixture is moulded, pressed and, after being wrapped in perforated tin foil, defatted by ether. It contains 7·2 percent moisture, 3·5 percent mineral substances, 89·3 percent organic substances, which are composed of nitrogenous substance 3·9 percent digestible albumin 18·8 percent, fat 18·3 percent, theobromine 0·46 percent, starch and other non-nitrogenous extractives 44·94 percent, cellulose 2·9 percent.[249] It is likewise a mixture of 50 percent of oat meat with 50 percent of cacao.

Oat-cocoa can be simply prepared by mixing cacao powder with an equal part of prepared oat meal, such as is produced by Hohenlohe's Präservefabrik, by Knorr of Heilbronn and by the Quaker Oats Company. In order to cover the taste of the cat meal 1-2 percent of sodium chloride is to be added.

J. Berlit, German patent No. 72449, describes the following method for the preparation of **oat-cacao**, Oats are cleaned, bruised, slightly roasted and ground. The powder is wetted and by means of a kneading machine worked up to a paste which is dried in a vacuum, finally ground and mixed with defatted cacao in the required proportions.

Palamoud des Turcs consists of cacao mass, rice-meal, starch and sandal wood.

Peptone-cocoa contains: water 4·08 percent, nitrogenous substance 20·56 percent, albumose 8·25 percent, peptone 4·41 percent, theobromine 1·03 percent, sugar 49·51 percent, besides non-nitrogenous constituents 9·37 percent, woody fibre 1·43 percent, mineral substance 4·17 percent (potash 1·97 percent, phosphoric acid 1·21 percent).

Peptone-powder-cocoa (20 percent) is prepared by mixing 20 parts of Koch's meat peptone in the form of extract with 50 parts of sugar and 40 parts cacao powder.

Peptone-chocolate contains 10 percent of dry peptone.

Plasmon-chocolate and cocoa contains 20 percent plasmon[250] (Siebold).

Racahout des Arabes see page 00, note.

Raspberry chocolate (Sarotti), German patent 181760 and 204603, prepared with addition of the juice of the raspberry.

Saccharin-cocoa gives the following results on analysis: water 7·26 percent, nitrogenous substance 20·5 percent, theobromine 2·09 percent, fat 32·25 percent, saccharin 0·4 percent, starch 13·02 percent, non-nitrogenous extractives 13·51 percent, woody fibre 5·27 percent, ash 5·93 percent, (potash 2·16 percent, phosphoric acid 1·69 percent). See also Hansa-Saccharin-cacao on page. 00.

Somatose-cocoa with sugar and somatose-chocolate contains about 10 percent somatose[251]; the first preparation contains 20·71 per cent total nitrogenous substance, and the latter 10·24 percent, of which about 1/3 consists of soluble nitrogenous compounds. (Mansfeld.) The first preparation could be readily prepared by mixing 10 parts of somatose (Farbwerke Bayer &. Cie., Elberfeld) with 50 parts of sugar and 40 parts of cocoa powder.

Theobromade (theobromine) is a dry extract from cacao husks.

442

Dr. Thesen's Proviant comes into commerce in the form of chocolate and is chocolate with an addition of albumin. Its analysis gives the following results: Albumin 20·5 percent, theobromine 0·56 percent, fat 39·79 percent, carbohydrates a) (soluble) 26·95 per cent, b) (insoluble) 5·66 percent, ash 2·25 percent, water 1·57 percent. A similar product to Thesen's Proviant results from mixing: albumin 12·5 parts, fat (cacao butter) 10 parts, fat (cocoa nut butter 7·5 parts, sugar 25 parts, cacao 45 parts.

Tropon-cocoa is a varying mixture of tropon, 15-33-1/3 percent, with cacao powder. A tropon cocoa containing 20 percent of tropon gave on analysis: water 5·75 percent, albumin 38·49 per cent., fat 27·77 percent, fibre 3·76 percent, ash 4·51 percent, theobromine 1·6 percent, extractives 22·78 percent.

Tropon-chocolate is a chocolate containing 25 percent tropon.[252]

Tropon-Oat-cocoa contains 20 percent of tropon, 30 percent of oat meal and 50 percent of cocoa powder.

Wacaca des Indes consists of 60 parts cacao powder, 165 parts sugar, 8 parts cinnamon, 2 parts vanilla and some tincture of ambergris.

White chocolate contains sugar 3000 parts, rice meal 860 parts, potato flour 250 parts, cacao butter 250 parts, gum arabic 125 parts and vanilla tincture 15 parts.[253]

A. Index to literature.

In the following list are specified in chronological order only those works and memoirs which refer to the culture of cacao and the manufacture of cacao preparations. The remaining literature on the subject, so far as it refers to the scientific side, has already been mentioned in the form of footnotes.

a) Cultivation.

Jumelle Henry, Le Cacaoyer, sa culture et son exploitation dans tous les pays de production, Paris 1900.

J. Hinchley Hart, F. L. S., Cacao, A treatise on the cultivation and curing of cacao. II. Edition. Trinidad 1900.

b) Technology.

Dictionnaire technologique ou nouveau Dictionnaire universel des arts et métiers et de l'économie industrielle par une société de savans et d'artistes. Paris 1823 et 1824. Tomes 4 et 5.

J. J. R. von Prechtl's Technologische Encyclopädie, Stuttgart Bd. III und Supplement-Bd. II. Stuttgart 1859.

Mitscherlich, A, Der Kakao und die Schokolade. Berlin 1859.

Zipperer, P, Die Neuerungen in der Fabrikation von Schokoladen und diesen verwandten diätetischen Produkten. Chemiker-Zeitung 1892 No. 58; 1893 No. 54; 1895 No. 21.

Gordian, A, Die deutsche Schokoladen-und Zuckerwaren-Industrie. Hamburg 1895.

Gordian, Zeitschrift für die Kakao-, Schokoladen und

Zuckerwaren-Industrie etc., Hamburg, seit 1896.

De Belfort de la Roque, L, Guide practique de la Fabrication du chocolat. Paris 1895.

Filsinger, F, Fortschritte in der Fabrikation von Schokolade und ihr verwandten diätetischen Präparaten in den Jahren 1895-1899, Chemiker-Zeitung 1897, No. 22 des Jahres 1897; ibid. 1898, No. 42 des Jahres 1898; des Jahres 1899, ibid. 1899, No. 48.

Spamer's, O, Buch der Erfindungen, Gewerbe und Industrieen. Leipzig 1897, Band IV.

Muspratt's Theoretische, praktische und analytische Chemie in Anwendung auf Künste und Gewerbe, begonnen von F. Stohmann und B. Kerl, herausgegeben von H. Bunte. Braunschweig 1898, Bd. VI.

Villon, A. M, Dictionnaire de Chimie industrielle, contenant les applications de la chimie à l'industrie, à la metallurgie, à l'agriculture, à la pharmacie, à la pyrotechnie et aux arts et métiers. Paris 1898, Tome premier.

Luegers, O, Lexikon der gesamten Technik und ihrer Hilfswissenschaften. Stuttgart und Leipzig 1899.

Ettling, Der Kakao, seine Kultur und Bereitung, Berlin 1903.

Kindt, Die Kultur des Kakaobaues und seine Schädlinge. Hamburg 1904.

Faber, Dr. F. C. von, Die Krankheiten und Parasiten des Kakaobaums. Berlin 1909. (Arb. aus der Kais. Biolog. Anstalt f. Landund Forstwissenschaft).

B. Tables.

Table

1. German Imports and Exports of cacao products 1907-1910 ... [35](35)

2. Imports in Germany 1900-1908 ... [37](37)

3. Imports or Consumption in the various countries ... [38](38)

4/5. Analysis of hulled bean ... [43/44](43/44)

6/7. Analysis of raw shelled bean (kernel) ... [44/45](44/45)

8. Analysis of Ridenour ... [45](45)

9. Analysis of roasted, shelled cacao (Matthes & Müller) ... [46](46)

10. Analysis of commoner varieties of cacao (Matthes & Müller) ... [47](47)

11. Analysis of cacao (defatted and free from alcali) ... [48](48)

12. Physical and chemical analysis of the various kinds of pressed Stollwerck Cacao Butter (Fritzsche) ... [56](56)

13. Constituents of different fats and oils contained in cacao ... [58](58)

14. Analysis of the ash of cacao beans by R. Bensemann ... [74](74)

15. Composition of cacao shells (Laube & Aldendorff) ... [76](76)

16. Analysis of unroasted cacao husks (Zipperer) ... [76](76)

17. Constituents contained in the ash of roasted cacao husks by R. Bensemann ... [77](77)

18. Fodder value of cacao husks (Maercker) ... [83](83)

19. Percentage of butter to be extracted ... [203](203)

Percentage of butter remaining in the finished

cacao powder

21. Adulteration and their detection   289

C. Illustrations.

Page

Fig. 1. Branch of cacao tree with blossom and leaves   2

Fig. 2. Fruit and single seeds in long and cross section   3

Fig. 3. Cross section of the cacao shell (enlarged)   14

Fig. 4. Cross section of edge of seed leaf (enlarged)   15

Fig. 5. Graph showing consumption of raw cacao   40/41

Fig. 6. Graph per head of population in Germany   42

Fig. 7. Grains and starch in cacao bean (section of ariba, enlarged 750 times)   70

Fig. 8. Plan of cacao shell (enlarged)   80

Fig. 9. Spongy paranchyma (enlarged)   80

Fig. 10. Dry cells or skereides (enlarged)   80

Fig. 11. Silver membrane with Mitscherlich particles (enlarged)   81

Fig. 12. Preliminary cleansing machine (J. M. Lehmann)   90

Fig. 13. Preliminary cleansing machine (J. M. Lehmann)   91

Fig. 13a. Brushing machine for cacao beans (Bauermeister)   92

Fig. 14. Cylindrical roasting machine (Lehmann)   93

Fig. 14. and b. Same in section   94/95

Fig. 15. a and b. Spherical safety roasters (Bauermeister)   96/97

Fig. 16.	Roaster with gas heating (Lehmann)	98
Fig. 17.	Cooling carriage with exhauster (Lehmann)	99
Fig. 18.	Crusher and cleanser (Lehmann)	101
Fig. 19.	Dust cleanser (Lehmann)	103
Fig. 20.	Electro-magnetic selecting machine (Lehmann)	104
Fig. 21.	Seed picking machine (Lehmann)	105
Fig. 22.	Seed picking (sectional drawing)	106
Fig. 23.	Simple cacao mill (Lehmann)	110
Fig. 24a.	Triple cacao mill (Lehmann)	111
Fig. 24b.	Triple cacao mill (Bauermeister)	112
Fig. 24c.	Triple cacao mill (Franke)	113
Fig. 25.	Fourfold cacao mill (Lehmann)	115
Fig. 26.	Cacao mill and roller apparatus combined (Bauermeister)	116
Fig. 27.	Warming through (Lehmann)	117
Fig. 28.	Preliminary mixing machine (Lehmann)	118
Fig. 29.	First melangeur (Hermann)	119

450

29.		[119](#)
Fig. 30.	Design of modern melangeur (Franke)	[121](#)
Fig. 31.	Modern melangeur with outlet at side (Lehmann)	[122](#)
Fig. 32.	Larger melangeur with cover and outlet (Lehmann)	[124](#)
Fig. 33a.	Design of first roller machine, front elevation (Savy)	[125](#)
Fig 33b.	do. Plan	[125](#)
Fig. 34.	Later machine (Savy)	[126](#)
Fig. 35.	Modern six roller machine by Lehmann	[127](#)
Fig. 36.	Nine roller apparatus (Bauermeister)	[128](#)
Fig. 37.	Same in design	[129](#)
Fig. 38.	Three roller machine with cast iron rollers (Lehmann)	[130](#)
Fig. 39.	2 three roller machines, attached to a "Battery"	[130](#)
Fig. 40.	Three roller machine with cast iron rollers (Franke)	[131](#)
Fig. 41.	and b. Four and five roller machines with cast iron rollers (Lehmann)	[132/3](#)
Fig. 41c.	Five roller machine with cast iron rollers (Bauermeister)	[134](#)
Fig. 42.	Three roller machine with electric motor (Lehmann)	[135](#)

Fig. 43.	Front elevation of triturating machine (Conche) by Franke	138
Fig. 43a.	Conche (Lehmann)	139
Fig. 44.	Conche room (Lehmann)	140
Fig. 45.	Warming closet with steam heating (Lehmann)	142
Fig. 46.	Small melangeur with one runner (Lehmann)	143
Fig. 47.	Do. modern construction (Lehmann)	144
Fig. 48.	Modern tempering machine (Lehmann)	145
Fig. 49.	Design of air exhausting machine (Lehmann)	147
Fig. 50.	Air exhausting machine (Lehmann)	147
Fig. 51.	and b. Chocolate dividing machines (Lehmann and Bauermeister)	148/9
Fig. 52.	Moulding and layering machine (Lehmann)	150
Fig. 53.	Reiche's mould cleansing and polishing machine	155
Fig. 54.	Design of shaking table	156
Fig. 55.	do, (Lehmann)	157
Fig. 56.	do, (Lehmann)	157
Fig.		

452

Fig. 57a.	do, Front elevation	159
Fig. 58,	58 a. and b. Shaking table batteries (Lehmann)	160/1
Fig. 59.	a. and b. Cooling plant (Wegelin & Hübner)	1655/6
Fig. 60.	do, perspective	167
Fig. 61.	Modern air cooling apparatus (Escher, Wyss & Co.)	169
Fig. 62.	Cooling plant of Cole's Arctic Patent Dry Cold Air Machine	170
Fig. 63.	a. and b. Cooling chambers by Lehmann	173/4
Fig. 63c.	Automatic moulding and cooling plant by Lehmann	175
Fig. 64/65.	Pastille machines (Reiche)	177
Fig. 66a-i.	Moulds to these machines	178
Fig. 67.	Pastille machines for thin chocolate material (Reiche)	179
Fig. 68.	Pastille and praliné metal hurdle (Reiche)	180
Fig. 69.	Mould metal Durabula Reiche	181
Fig. 70.	Fondant machine (Lehmann)	183
Fig. 71.	do, modern construction (Lehmann)	184
Fig.	Fondant casting machine (Lehmann)	

Fig. 72.	Fondant casting machine (Lehmann)	185
Fig. 73.	Fondant powdering off machine, for hurdles (Lehmann)	186
Fig. 74.	do, non-stop (Lehmann)	187
Fig. 75.	Coating machine (Lehmann)	188
Fig. 76.	Stirring apparatus for coating material (Lehmann)	188
Fig. 77/78.	Coating or dipping machines (Reiche)	189
Fig. 79/80.	Grating to these	190/1
Fig. 81.	Modern dipping machine constructed by Lehmann	193
Fig. 82.	Cacao press, 400 atmospheres (Lehmann)	201
Fig. 83a.	Cacao butter filter, design (Hänig & Co.)	202
Fig. 84.	Cacao press on larger scale (Lehmann)	205
Fig. 84a.	Pump for cacao press (Lehmann)	206
Fig. 84b.	Cacao cake crusher (Seek)	207
Fig. 85a.	do, (Bauermeister)	208
Fig. 85b.	do, (Lehmann)	209
Fig.	Pulveriser (Lehmann)	

454

Fig. 87.	Pulverising and sifting machine (Lehmann)	211
Fig. 88.	Centrifugal sifting machine, modern construction (Lehmann)	213
Fig. 89.	Automatic pulverising plant (Lehmann)	215
Fig. 90a.	Mixing machine (Lehmann)	217
Fig. 90b.	Universal kneading and mixing machine (Werner & Pfleiderer)	218
Fig. 91a.	Vacuum kneader, closed (Werner & Pfleiderer)	219
Fig. 91b.	Vacuum kneader, open and upturned (Werner & Pfleiderer)	221
Fig. 92.	Filling and packing machine	229
Fig. 93.	do, "Triumph" (Fritz Kilian)	229
Fig. 94.	Edge-runner mill	231
Fig. 95.	Drum sifting machine (Lehmann)	232
Fig. 96.	Combined sugar-grinding and sifting apparatus (Lehmann)	233
Fig. 97.	Spice and stamping apparatus (Lehmann)	239
Fig. 98.	Pulverising mill (Savy)	240
Fig. 99.	Sifting machine (Savy)	241
Fig.	Parenchyma of the cotyledon, enlarged	

455

Fig. 100.	Parenchyma of the cotyledon, enlarged	275
Fig. 101.	Cocoa powder, enlarged	276
Fig. 102.	do, enlarged	277
Plate I.	The Cacao Tree	
Plate II:	Chocolate factory (design)	305
Plate III:	Cocoa powder factory (design)	306

D. Authors. Alphabetical index.

Abels, Franz, 313

Albanese, 64

Aldendorf & Laube, 44, 76, 77

Allihn, F., 265

Altschul, J., 244

Abt, J., 311

Arning, 243

Aufrecht, 310, 311

Baier, 272, 273

Bastin, E. S., 70

Baudrimont & Chevalier, 50

Baudonin, 59

Bauermeister, H., 148, 214

Bayer & Co., 317

Beam & Leffmann, 276

Beckurts, H., 74, 228, 236, 261

Beddies, Alfr., 316

Benedict, 55, 57, 59, 261

Bensemann, R., 74, 77

Berg & Schmidt, 2, 3

Berger, Th., 153

Berlit, J., 316

Beythien, 277

Bilterist, 272

Björklund, 53, 261, 262

Boehme, Dr. Rich., 258, 285

Börnstein, 236

Bondzynski & Gottlieb, 64

Bonnema, 245

Bonteköe, 6

Bordas & Touplain, 271

Bourot & Jean, 59

Boussignault, 83

Bozelli, 85

Branlatio, 6

Brissemoret, 65

Buchat, 6

Buisson, 85

Burstyn, 53, 54, 55

Busse, W., 241, 243

Carletti, Antonio, 6

Chalot, C., 7

Charles V., 5

Chevalier & Baudrimont, 50

Cibil, 314

Clusius, 6

Cohn, 56, 260

Cole, 170

Cortez, Fernando, 5

David & Söhne, 271

Dekker, 65, 267

Denayer, A., 308

Desprez, 199

Dieterich, E., 307, 308, 313, 314

Dietrich, K., 51, 249, 250

Dingler, 53

Donat, von, 311

Dove, 85

Dowson, 97

Dragendorff, 66

Drave, 267, 268

Ducleaux, 75

Eminger, 65, 263, 264

Escher Wyss & Co., 168

Ester, 264

d'Estrées, 6

Ettling, K., 320

Faber, Dr. F. C von, 8, 87, 320

Faelli, Prof., 83

Fahlberg, 234

Farnsteiner, 255, 256, 258

Fehling, 71, 237,, 271

Filsinger, 44, 52, 53, 54, 57, 72, 81, 83, 107, 261, 267, 268, 269, 288, 299,, 319

Filsinger & Henking, 52

Fischer, B. & Grünhagen, 267

Fischer, Emil, 63, 68

Forster, 69

Franke, Paul & Co., 233

Fresenius, C. R., 256

Freudenberg, Ph., 4

Freudenberg, W., 5

Fritzsche, Dr., 55

Gädke, 225

Galippe, 75

Gérard, 264

Gieseler, 243

Goethe, J. W., 16

Gordian, 87, 319

Gottlieb & Bondzynski, 64

Graf, 50

Gram, Chr., 64

Greiert, 39

Greiner, 270

Groult & Boutron-Russel, 311

Grünhagen, B. & Fischer, 267

Gruson, 239

Guenez, E., 277

Guerin, 243

Haarmann, W., 244

Haarmann & Reimer, 244

Hänig, Volkmar & Co., 202

Härtel, 273

Hager, 51, 312

Hahn-Holfert, 307, 312

Hanausek, J. V., 12, 238

Hausen & Co., 312, 316

Hart, J. Hinchley, 10

Hartwig & Vogel, 307

Haubold, C. G., 164

Hauswaldt, W., 87, 88

Hefelmann, 245

Heisch, C., 44

Henking & Filsinger, 52

Henneberg, 72, 267

Henning, 246

Hensel, Dr. & Co., 222

Hermann, G., 86, 120, 123, 126

Hess & Prescott, 245

Hesse, William, 243

Hilger, 11, 60, 61

Hilger & Lazarus, 61

Hockauf, 74

Hohenlohe, 316

van Houten, C. J., 59, 195

von Hübl, 53, 56

Husson, 312

Jean & Bourot, 59
Jeserich, 269

Kathreiner, 82
Keller, C. C., 65
Kilian, Fritz, 229
Kindt, L., 9, 10, 320
Kingzett, 50
Kjeldahl, 171
Klimont, 50
Knorr, 316
Knoch, 317
Koeben, Dr., 315
König, 72, 76, 266, 267
Köttsdorfer, 54
Kreplin, E., 312
Krupp, 125, 239

Lahmann, 315
Lagerheim, G., 267
de Laire, G., 244
Lampadius, 43
Laube & Aldendorff, 44, 76, 77
Laxa, 273, 274
Lazarus & Hilgers, 61
Leffmann & Beam, 277

Lehmann, Berlin, 309

Lehmann, J. M., 100, 105, 121, 132, 148, 172, 202, 210, 233

Létang, 154

Lewkowitsch, 50, 51, 262

Leys, 271

L'Hôte, 74

Liebig, 314

Linné, 6

Lobeck & Co., 224

Loher, 11

Louis XIV., 182

Louis XVI., 6

Lueger, O., 320

Lührig, H., 267, 271

Macquer, 199

Maerker, 82

Majert & Ebers, 311

Mansfeld, 317

Matthes, 72

Matthes & Fritz Müller, 45, 74, 77

Maupy, 66

Mayfarth, 10

Meissl-Reichert, 53, 55, 260, 273, 274

Merck, E., 252, 264

Mering, 313

Merz, 54

Meyer-Finkenburg, 302

Michaelis, 306

Michel, Alfr., 82

Mitscherlich, A., 5, 13, 43, 63, 85, 92, 126, 199, 276

Moeller, 15, 16, 79, 237, 241

Molisch, 16, 67, 75

Moser & Co., 216

Müller, Matthes & Fritz, 45, 74, 77

Muspratt, 120, 126, 320

Nencki, L., 235

Neumann, R. O., 203, 226

Notnagel, 310

Oldam & Withe, 52

Onfroy, P., 255

Paris, G., 76

Payen, 43

Peckoldt, Th., 12

Pelletier, 86

Petzholdt, J. S., 148

Pieper, 198

Pintus, 109

du Plessis, 182

Polenske, 56

Posetto, 266

Pralin, 182

von Prechtl, 319

Prescott & Hess, 245

Preyer, Dr. A. von, 11

Py, 277

Rammsberger, 51

Rauch, F., 6

Reichardt, 309

Reiche, Anton, 119, 152, 153, 178, 182, 189

Reichert-Meissl, 53, 55, 260, 273, 274

Reinhardt, G., 88

Ridenour, 44, 45

Riederer, 264

Rimbach, Dr. C., 13

Riquet & Co., 307

Rocques, 55

Roque, Belfort de la, 55, 319

Rost, 64

Rouché, 244

Royer, 228

Rüger, Otto, 217, 224

Ruffin, A., 57

Savy, A. & Co., 228

Sarotti, 317

Schimper, A. F. W., 13, 16

Schmidt, <u>51</u>

Schmidt & Berg, <u>2</u>, <u>3</u>

Schrader, <u>62</u>

Scholz, J., <u>309</u>

Schröder, W. von, <u>64</u>

Schütte-Felsche, Wilh., <u>102</u>

Schweitzer, C., <u>11</u>, <u>60</u>, <u>61</u>,, <u>73</u>

Seck, Gebr., <u>207</u>

See, G., <u>64</u>

Sévigné, Madame de, <u>6</u>

Sieberts, <u>309</u>

Siebold, <u>317</u>

Skalweit, <u>75</u>

Soltsien, <u>71</u>

Soxleth, <u>259</u>, <u>264</u>, <u>273</u>

Spamer, O., <u>320</u>

Spindler, <u>315</u>

Stähle, C., <u>197</u>

Steinmann, A., <u>271</u>

Stollwerck, Dr. W., <u>34</u>

Stollwerck Broth., <u>56</u>, <u>88</u>, <u>258</u>

Strecker, <u>62</u>

Streitberger, <u>72</u>

Strohl, <u>54</u>

Strohschein, <u>82</u>

Stutzer, A., <u>69</u>

Suringar & Tollens, 72, 267

Theinhardt, Dr., 312

Theresia of Austria, 6

Thesen, Dr., 317

Thiele & Holzhause, 312

Timpe, Th., 307

Tollens & Suringar, 72, 267

Touplain & Bordas, 271

Trojanowsky, 74

Tschirch, 14

Tuchen, 74

Ulzer, Benedict-, 57, 261

Villon, 320

Villon-Guichard, 169

Wagner, L., 229

Weender, 72, 108, 266

Wegelin & Hübner, 165

Weldon, 109

Welmans, 5, 53, 55, 243, 258, 260, 266, 268, 269, 271, 275, 276, 277

Wendt, G., 198

Werner & Pfleiderer, 137, 219

White, 51

White & Oldam, 52

William, Prince of Lippe, 6

William of Brandenbourg, 6

Wolfram, 66

Woseressenzky, 62

Woy, Rud., 270, 271

Zeiss, 55, 261

Zipperer, 16, 44, 52, 74, 76, 77, 83, 229, 270

E. Index.

Accra-Cacao, 17, 29

Acid benzoic, 243

Acid hydrochloric, 16

Acid yellow, 251

Acids, solid, fatty, 53

Acids volatile, 53

Acids, sugar and plant—, 73

Acid value, determination of, 54

Acorn-Cacao, Michaelis, 306

Acorn-Cacao, Hartwig & Vogel, 307

Acorn-Cacao, Th. Timpe, 307

Acorn-Chocolate, 307

Acorn-Malt-Cacao, Dieterich, 307

Acorn-Malt-Chocolate, 307

Acrolein, formation of, 50, 93

Adraganth, 255

Adulteration of cocoa goods and its detection, 288

African cacao varieties, 28

Air, removal of, 143

Air extracting machines, 144

Albumin, 67

Albuminates, determination of, 271

Albuminous chocolate and cocoa, 307

Albumoses, 67

Alcohol ether test, Filsinger's, 262

Aleuron granules, 67

Alizarin blue, 251

Alkali solution, 222, 224

Alkalis for soluble cocoa, 196, 216, 222

Alkalis fixed, 198

Alkalis remaining in the cocoa, estimation of, 256

Alkaloids, 63

Amaranth, 251

American cacao varieties, 19

Ammonia, 164

Analysis of cacao, 48

Analysis of cacao-butter, 58

Analysis of mixtures of different blends, 109

Analysis of the raw shelled bean, 44, 45

Analysis of the various kinds of pressed Stollwerck cocoa butter, 56

Analysis of waste products, 108

Analysis and examination of cocoa preparations, 253

Anilin blue, 251

Anilin colours permissible, 250

Antifebrin, 245

Aroma of the bean, 59

Arriba cacao, 17, 20

Arrowroot, 237

Arctic Machines, Cole's, 170

Artificial refrigeration, 163

Ash, estimation of, 255

Ash or mineral constituents, 73

Ash remaining in raw and shelled cacao beans, 74

Asiatic cacao varieties, 32

Aspergillus, 242

Australian cacao varieties, 33

Automatic dividing machines, 146

Automatic filling and packing machine 229

Bahia Cacao, 22

Bahia de Caraquez, 21

Balao, 21

Barley-Chocolate, 308

Battery-Refiners, 132

Battery-Shaking Tables, 160

Beans, in general, 1

Beans, description of, 12

Beans, preliminary treatment of, 197

Beans, preparations of, 85

Bean meal, 238

Benzoic acid, 243

Benzoic tincture, 243

Benzoin, gum-, 249

Björklund's ether test, 262

Bordeaux red, 251

Botanical definition of the cacao tree, 5

Brazil cacao, 22

Brilliant blue, 251

Brine for cooling purposes, 165

Brushing machine for cacao beans, 89

Burning of chocolate mass, avoiding it, 134

Butter of Cocoa, 58, 138, 187, 195, 284, 286

Buttneriaceae, 5

Butyro-refractometer, 55

By-products in the cocoa industry, 81

Cacaohoatel, 5

Cacao beans, 1

Cacao beans, description of, 12

Cacao beans, preparation of, 85

Cacao beans, preliminary treatment of, 197

Cacao blanco, 13

Cacao butter, 58, 138, 187, 195, 284, 286

Cacao butter filters, 202

Cacao butter, percentage to be extracted, 203

Cacao butter, remaining in the finished cocoa, 204

Cacao cake crusher, 210

Cacao egg-cream, 308

Cacao essence, 308

Cacao fruit and flowers, 1, 2

Cacao glycoside, 60

Cacao husk, determination of, 267

473

Cacao liqueur, 308

Cacao, malt, 308

Cacao mass, production of, 109, 282, 285

Cacao mills, 110

Cacao plantation, 7

Cacao powder, 105, 187, 195, 210, 282, 285, 290

Cacao powder-factory, installation of, 306

Cacao preparations, definition of, 279

Cacao presses, 199

Cacao red, 43, 59

Cacao shells, 1, 2, 76, 82

Cacao soluble, 105, 195

Cacao, substances of, 49

Cacao tincture, 310

Cacao tree, cultivation, diseases and parasites, 7

Cacao tree, description of, 1

Cacao tree, distribution and history, 4

Cacao and chocolate preparations containing milk, 308

Cacaol, 308

Cacaophen Sieberts, 309

Cacap, 5

Cacava-quahitl, 5

Cacogna, 195

Caesalpina, 7

Caffeine, determination of, 263

Caracas, 17, 25

Caraquez, 21

Carbonic acid for cooling purposes, 164

Cardamoms, 248

Cardamom oil, 249

Carob in the cacao, 278

Carupano cacao, 25

Castilloa, 7

Cauca bean, 20

Cellulose or crude fibre, 72, 266

Centrifugal sifting machine, 210

Ceylon Cacao, 5, 32

Chemical and microscopical examination of cocoa preparations, 253

Chemical constitution of the bean, 43

Chestnut meal, 238

Children's Nährpulver, 309

Chilled metal rollers, 125, 130

Choclean, 309

Chocolate, manufacture of, 85, 283

Chocolate-cigars, 152

Chocolate cooling plants, 166

Chocolate cream syrup, 309

Chocolate croquettes, 181

Chocolate, crumb-, 153

Chocolate digestif, 309

Chocolate, dividing it, 143

Chocolate eggs, 153

Chocolate factory, installation of, 305

Chocolate, Fondants-, 138, 189

Chocolate, health-beer-, 309

Chocolate, hygienic, 136

Chocolate lozenges and pastilles, 176

Chocolate, milk-, 141, 222, 272, 284, 286

Chocolate, moulding it, 150

Chocolate moulds various, 151, 152, 153, 154

Chocolate powder, 283, 286

Chocolate raw, treatment of, 138

Chocolate rétablière, 309

Chocolate spiced, 136

Chocolate syrup, 310

Chocolate tincture (cacao tincture), 310

Chocolate vanilla, 136

Chocolate varnish, 250

Chocolatl, 5

Christmas tree articles, 181

Cinchona red, 60

Cinnamon, 246

Cinnamon oil, 249

Cleaning machine for moulds, 154

Cleaning machine for beans, 90, 91

Cleaning, storing and sorting of the beans, 87

Cloves, 247

Clove oil, 249

Coated chocolates, 182, 187

Coating materials, 138, 141, 182, 187, 283, 286, 310

Coffie-mama, 7

Cole's Arctic Machines, 170

Colour of the cotyledon, 9

Colouring of cocoa powder, 204

Colouring materials, 250

Coloration of starch with iodine, 71

Columbia, 19

Combined cocoa mill and refiner, 116

Commercial kinds of cacao, 12, 16

Commercial value of raw cacao, 17

Compressor, 164

Composition of the hulled bean, 43

Conches, 138

Condenser, 155

Constituents, mineral or ash-, 73

Constituents of cacao husks, 76

Constituents in ash of cacao husks, 77

Constitution of the bean, chemical, 43

Consumption of cocoa products, 33, 38, 42

Consumption of coffee, cocoa and tea, comparison, 39

Cooling cellars, 168

Cooling chambers, 152

Cooling the chocolate, 162

Cooling the roasted beans, 100

Cooling trucks with exhaust apparatus, 100

Copper in the ash of beans and husks, 75

Coriander oil, 249

Corn cacao, 310

Costa Rica, 19

Cotyledon, 15

Covering or coating materials, 138, 141, 182, 187, 283, 286,, 310

Cream chocolate, examination of, 272, 284

Criollo, 18

Crude fibre, 72, 266

Crumb chocolate, 153

Crushing of cocoa and sugar lumps, 122, 210

Crushing, hulling and cleaning of the beans, 100

Crushing, hulling and cleaning machines, 101

Crystal sugar, 231

Cuba, 28

Cultivation of the cacao tree, 7

Cumarin, 244, 245

Declaration of added ingredients, 281

Defatted cocoa, 203, 208

Definitions of cocoa preparations, 279

Depositing machine, 186

Description of the beans, 12

Dextrin, 237

Dextrose, 71, 265

Diabetic chocolate, 310

Diabetic cocoa, 311

Dictamnia, 311

Dietetic cocoa preparations, 306

Diorit rollers, 125

Dipping machine, 192

Dipping of pralinés, 137, 189

Diseases of the cocoa tree, 7

Disintegrating the cocoa tissues, 195

Disintegration, methods of, 197

Disintegration before roasting, 197

Disintegration after roasting, 216

Disintegration prior to pressing, 217

Disintegration after pressing, 224

Disintegrators, 233

Distribution of the cacao tree, 4

Diureides, 62

Diuretin, 64

Dividing machines, 148, 149

Division of chocolate, 143

v. Donat's albumin chocolate, 311

Double cocoa mills, 114

Dowson gas, 97

Dry cocoas, 208

479

Dulcin, 235

Durabula-moulds, 182

Dust particles in cacao beans, 102

Dutch cocoas, 195, 203

Dutch IIa cocoa butter, 82

Earth nut in the cocoa, 278

Easin, 251

Ecuador, 20

Electric motors, 134, 168

Electro-magnetic metal extracting machine, 103

Erythrina indica, 7, 8

Erythrosin, 251

Esmeraldas, 22

Estates, 26

Estimation of alkalis remaining in the cocoa powder, 256

Estimation of albuminates, 271

Estimation of ash, 255

Estimation of cocoa husk, 267

Estimation of crude fibre, 266

Estimation of the fatty contents, 258

Estimation of moisture, 254

Estimation of silicic acid in the ash, 256

Estimation of starch, 264

Estimation of theobromine and caffeine, 263

Ether oils, 248

Ether test, Björklund's, 262

Eucasin chocolate and cocoa, 311

Evaporator, 164

Examination and analysis of cocoa preparations, 253

Exports from Germany, 35

Extraction of cocoa butter, 195, 199, 203, 204

Fair shipping cocoa, 26

Fat contained in cocoa, 49

Fat contained in cocoa shells, 57

Fat, extraction of, 195, 199, 203, 204

Fatty contents, determination of, 258

Fermentation of the beans, 9, 60, 198

Fermentation secondary, 87

Fermentation tanks, 10

Fernando Po, 32

Fibre, determination of, 108

Fibre crude, 72, 254, 266

Fibre woody, 108

Filsinger's alcohol ether test, 262

Filters for cocoa butter, 202

Flavour of the finished cocoa powder, 206, 226

Flavouring matter (spices), 287

Flour, 236

Fodder value of the husks, 83

Fondant chocolate, 138, 182

Fondant machines, 183, 184

Food salt cocoa, 315

Food and health powder, 315

Forastero, 19

Fuchsin, 251

Galactogen cocoa, 312

Gathering and fermentation of the beans, 9

Gauga, 312

Gelatine, 255

Geographical distribution and history of the cacao tree, 4

Germ separating machine, 105

Globoids, 75, 276

Globulins, 68

Glucin, 235

Glucose, 71, 138

Glycoside, 11, 60, 253

Gold Coast, 28

Granite rollers, 123

Granulated sugar, 231

Grinding and trituration of the cocoa mass, 109

Guadeloupe cacaos, 26

Guarana paste, 16

Guayaquil cacaos, 17, 20

Guiana, 23

Gum benzoin, 249

Gum disease, 8

Haema chocolate, 312

Haiti cacaos, 27

Hansa saccharin cocoa, 312

Hardidalik, 312

Hazelnut pulp in cocoa, 278

Heating of the cocoa mass, 117

Heating trough, 117

Heating chambers and closets, 141, 142

Heliotropium, 242

Hensel's Nähr-cacao, 312

Hetero albumose, 68

Hetero xanthine, 64

History of the cacao tree, 4

Homeopathic chocolate, 312

Hulled bean, composition of, 43

Hulling the cacao beans, 100

Husks of cocoa, 76, 82, 267

Husks, fodder value of, 83

Husson's mixture, 312

Hydraulic presses, 199

Hygiama, 312

Hygienic chocolate, proportions for mixing it, 136

Iceland moss chocolate, 313

Imports to Germany, 35, 37

Imports or consumption in the various countries, 38

Index, refractive-, 55

Indigo, 60

Indigosulfone, 251

Induline, 251

Ingredients added, declaration of, 281

Ingredients condemned, 230

Ingredients used for chocolate, 230

Iodine value, 53, 54

Java cacao, 17, 33

Kaiffa, 313

Kameroon cacaos, 19, 29

Kernels, analysis of, 44, 45, 76

Kneading and mixing machines, 217

Kola chocolate, 313

Kola nut, 60

Kongo, 30

Kraft chocolate, 313

Lagos, 29

Leguminous meals, 238

Levigation of chocolate, 81, 123

Lipanin chocolate, 313

Loss of weight by roasting, 96

Lozenges, 176

Mace, 247

Mace oil, 249

Machalla, 20

Malachite green, 251

Malt cacao, 313

Malt cacao-syrup or malted chocolate, 313

Malt chocolate, 313

Malt extract-chocolate, 313

Malto-leguminose cacao, 313

Manioc, 7

Manufacture of cocoa powder and soluble cocoa, 195

Manufacture of cocoa preparations 85, 282

Manufacture of chocolate, 85, 283

Maracaibo, 25

Martinique cacaos, 26

Meat-extract-chocolate, 314

Melangeurs, 121, 122, 124, 209,, 217

Melting kettle, 187, 188

Melting point of the cocoa butter, 52, 117, 261

Methylviolet, 251

Mexican cacaos, 19

Microscopic-botanical investigation, 275

Microscopic-chemical examination of cocoa preparations, 253

Milk chocolate, manufacture of, 141, 222, 286, 314

Milk cocoa, 314

Milk and cream chocolate, examination of, 272, 284

A more bitter milk cocoa, 314

Milk cocoa sweet, 314

Mill and refiner combined, 116

Mineral or ash constituents, 73

Mitscherlich particles, 13

Mixing cocoa powder with alkalis, 223

Mixing different kinds of cocoa, 108, 109

Mixing machines, 118, 210, 217

Mixture with sugar and spices, 117

Moisture, contained in cocoa, 49

Moisture in cocoa powder, 222

Moisture, estimation of, 254

Monomethyl xanthine, 64

Motors, electric, 134, 168

Moulds, 151, 152

Mould cleaning machines, 154

Moulding the chocolate, 149

Moulding machines, 150

Mucor circinelloids, 242

Murexide reaction, 66

Mutase-cacao, 315

Mutase-chocolate, 315

Nährsalz-cacao (Lahmann), 315

Nähr- und Heilpulver, 315

Naphtolyellow, 251

Naranjal, 21

Natural cocoa and chocolate, 315

Nicaragua cacao, 19

Nips, 11

Nuco-cacao, 315

Nutmeg, 247

Nutmeg oil, 249

Oat-cocoa Berlit, 316

Oat-cocoa Hallenser, 316

Oat-cocoa Kasseler, 316

Official enactments respecting the trade in cocoa preparations, 280

Official enactments respecting the trade in cocoa preparations Belgium, 291

— Roumania], 293

— Switzerland, 294

— Austria, 298

— Germany, 301

Oidium of cocoa, 228

Oils, ether-, 248

Oil sugar, 249

Opening up the cacao tissues, 195

Orange I, 251

Orange L, 251

Ornamented goods, 181, 189

Oscuros, 21

Packet filling machine, 228

Packing and storing of finished cocoa preparations, 227

Palamoud des Turcs, 316

Para cacao, 23

Parasites of the cacao tree, 7

Pastilles, 176

Pastille machines, 177, 179

Paternoster, 192

Pegados, 21

Pelatos, 21

Peptons, 68

Peptone-cocoa, 316

Peptone-chocolate, 317

Peptone-powder-cocoa, 317

Percentage of butter to be extracted, 203

Percentage of butter remaining in the finished cocoa, 204

Peru, 22

Peru balsam, 249

Peruviol, 249

Phloxin, 123

Pigment, 59

Plansieves for cocoa powder, 214

Plantation, 26

Plasmon chocolate and cocoa, 317

Polen's value, 260

Ponceau red, 251

Porcelain rollers, 215, 133

Porphyry rollers, 123

Potato starch, 236

Powder, chocolate-, 283

Pralinés, 182, 187, 189

Preliminary crushers, 212

Preparation of the cacao beans, 85

Presses, hydraulic-, 199

Production of the cocoa mass, 109

Proportions for mixing cocoa mass, sugar and spices,

136

Proteins, 67

Proteoses, 68

Puerto Cabello, 25

Pulverisation of the cocoa, 195

Pulverisation of the seeds, 199

Pulverisers, 210, 233, 239

Pulverising plant, 211, 212

Pulverising and sifting the defatted cocoa, 209

Pulverising the sugar, 233

Quadruple cocoa mills, 115 348

Racahout des Arabes, 317

Raspberry chocolate, 317

Raw fibre, 254

Raw shelled bean (kernel) analysis of, 44, 45

Refining machines (rollers), 126, 134

Refiner and mill combined, 116

Refractive index, 55

Refractometer-butyro, 55

Refrigeration, artificial-, 163

Reichert-Meissl value, 55

Removal of air and division of the chocolate, 143

Rice starch, 237

Roasting the cacao beans, 89, 199

Roasting machines, 93

Root bark of cacao, the use of it, 11

Roscellin, 251

Saccharin, 234

Saccharin-cocoa, 317

Salep, 238

Samana, 17

Samoa, 33

San Antonio, 26

San Thomas, 30

Sanchez, 17, 27

Santo Domingo, 27

Saponification of cocoa fat, 53, 54

Secondary fermenting, 87

Seed membrane of the bean, 11, 15

Semi-dipped goods, 192

Shaking tables, 156

Shaking table-batteries, 160

Shellac bleached, 250

Shell of the cacao bean, 14, 76

Shelling of the cacao beans, 100

Sifting the defatted cocoa, 209

Sifting machines, 210, 232

Silicic acid in the ash of cocoa, 256

Silver membrane, 79

Simple cocoa mills, 110

Soconusco, 26

Soluble cocoa, 105, 195

Somatose-cocoa with sugar, 317

Somatose-chocolate, 317

Spices and sugar, 117, 238, 287

Spiced chocolate, proportions for mixing it, 136 349

Starch cleaning machines, 186, 187

Starch, coloration of, with iodine, 71

Starch determination of, 264, 277

Starch foreign in cocoa, 275

Starch granules, 16, 70, 275

Starch, kinds of, 236

Starch powder, 185

Starch sugar, 71

Statistics of the cocoa trade, 35

Steel rollers, 125, 130

Stirring machines, 187

Storing and packing of finished cocoa preparations, 227

Storing and sorting of the beans, 87

Substances albuminous, 67

Substances occurring] in cacao, 49

Sucramin, 235

Sugar, determination of, 269

Sugar and plant acids, 73

Sugar and spices, 117, 231, 238, 287

Sugar, boiling it, 183

Sugar dust, 231

Sugar flour, 231

Sugar pulverising machines, 233

Sugar sifting machines, 232

Suisse Fondant machines, 138

Surinam cacao, 23

Sweetmeats, 186

Sweetening stuffs, 231

Sweets laquer, 250

Sykorin, 234

Sykose, 234

Syrup, 183

Temperature in cooling chambers, 172

Temperature in heating chambers, 141

Temperature for chocolate fondant and milk chocolate, 141

Temperature for moulding chocolate, 150

Temperature for roasting the beans, 89

Tempering machines, 144, 145, 188

Tenguel, 21

Testing the cocoa powder and chocolate, 253

Theobroma cacao, 5, 12

Theobromade, 317

Theobromine, 16, 43, 62, 263

Dr. Thesen's Proviant, 317

St. Thomas, 30

Tin boxes, 227

Tincture of benzoin, 243

Togo, 29

Trade in cocoa, 32

Tragacanth in cocoa goods, 277

Treatment of the cocoa mixture, 119

Trinidad-Criollo, 26, 32

Triple cocoa mills, 111

Trituration of the cocoa mass, 109, 119

Tropaedlin, 251

Tropon-cocoa, 317

Tropon-chocolate, 318

Tropon-oat-cocoa, 318

Trough, heating-, 117

Tumaco-cacaos, 20

Ureides, 62

Uropherin, 64

Vacuum kneader, 220

Vanilla, 241

Vanilla-chocolate, proportions for mixing it, 136

Vanillin, 119, 241, 243

Vascular bundles, 16

Venezuelan cacao, 17, 24

Volatile acids, 53

Wacaca des Indes, 318

Walnut pulp in the cocoa, 278

Waste products in cleaning, 106

Waste products in sifting, 107

Waste products in sorting, roasting, crushing and hulling, 107

Water blue, 251

Water cooling of steel rollers, 131

Water or moisture contained in the cacao, 49, 254

Weighing-machines, 148, 149

Wheat starch, 236

White chocolate, 318

Woody fibre, 108

Yellow acid R, 251

Zuckerin, 234

FOOTNOTES:

1 Of which the Central Province has 32,003 acres: North Western Province 3689 acres, North Central Province 25 acres, Province of Uva 2153 acres and Province of Sabaragamura 1918 acres. (From information kindly furnished in a letter of W. Freudenberg jun. German Consul at Colombo.)

2 See references at the end of this book.

3 Pronounced Chocolatl.

4 Revue des sciences pures et appliquées 1899, No. 4, page 127.

5 Vol. 7, Part 2: Diseases and Parasites of the Cacao Tree. With special reference to the conditions obtaining in the colonies belonging to Germany. By Dr. F. C. Faber, Berlin 1909, Parey & Springer.

6 Recently so-called fermenting-houses, as recommended by L. Kindt. (Cf. Kultur d. Kakaobaues und seine Schädlinge, Hambourg 1904), have answered very well. Yet the chemismus of fermentation is by no means sufficiently explained, and quantitatively and qualitatively, there is a lack of completeness in the analyses bearing on the process.

7 Special ovens (System Mayfarth) are also used, and sometimes complete heating and drying installations.

8 This had already been noticed by J. Hinchley Hart; Cacao (Trinidad 1892). It is therefore scarcely conceivable that the "Germination" theory should have held the field so long.

9 According to Schweizer (Pharmazeut. Ztg. 1898, page 389) these substances would be represented by the chemical formula $C_{60}H_{86}O_{15}N_4$, corresponding to 1 molecule cacao red, 6 molecules grape sugar, and 1 molecule Theobromin.

10 Cf. Hilger, Apotheker-Ztg. 1892, p. 469.

11 Cf. Tropenpflanzer V. 4, 1901, April-Number.

12 Loc. cit. page 167.

13 The leaves of the tobacco plant must also be fermented, before they acquire their rich brown colour and peculiar aroma.

14 Reports of the German Pharmaceutical Society 1900, Vol. 5, page 115.

15 J. F. Hanousek, Die Nahrungs-und Genußmittel aus dem Pflanzenreiche. p. 437.

16 Anleitung zur mikroskopischen Untersuchung der Nahrungs-und Genußmittel. Jena 1886.

17 Grundriß einer Histochemie der pflanzlichen Genussmittel.

18 See page 16 loc. cit.

19 Cf. Dr. Stollwerck. The Cacao and Chocolate Industries.

20 Mitscherlich, p. 57.

21 Cacao and its Preparation; a few Experiments.

22 Ridenour, M. American Journal of Pharmacy, 1895. Vol. 67, p. 207.

23 Filsinger, Chemical Journal, 1887, p. 202.

24 Z. U. N. G., 1906. Vol. 12, p. 88 et seq.

25 The husks contain no fat when in a fresh condition but absorb fat from the bean when the cacao is fermented and dried; especially so also in the later process of roasting, when they become saturated with it.

26 Klimont, Ber. d. Dtsch. chem. Ges. 34, 2636; Monatssch. f. Chem. 1902 (23) 51; 1904 (25) 929; 1905 (26) 536.

27 Journal of the Society of Chemical Industry 1899, p. 556.

28 Chevalier & Baudrimont, Dictionnaire des alterations.

29 Achiv de Pharmacie 1888, Vol. 26, p. 830.

30 See previous reference.

31 Schmidt, Ztschr. analyt. Chem. 1898, vol. 301 p. 301; cf. also P. Welmans, Pharm. Ztg. 1894, p. 776.

32 Pharm. Zeitung 1898 No. 10.

33 Cor. Assoc. Germ. Choc. Man. 1889, Vol. 5, p. 65.

34 The Brit. and Colon. Druggist 1897 No. 21.

35 Zeitschr. anal. Chemie

36 The Reichert-Meissl number (to be discussed later), according to a communication from P. Welmans, reaches 1 Burstyn in the expressed fat and amounts to 1·66 cc. in the extracted fat (no. of cc. of normal potash solution to 100 grammes of fat).

37 Dingler, Polytechnical Journal, Vol. 253, p. 281. For details of the method compare also P. Welmans Zeitschrift für öffentl. Chemie, 1900, No. 5.

38 Zeitschrift für anal. Chemie 1896, p. 519.

39 Zeitschrift für öffentl. Chemie 1900, p. 95.

40 Though Strohl Zeit. Analyt. Ch. 1896. Vol. 35. p. 166. has obtained with a Bahia fat an iodine value of 41·7, possibly exception due to some over-roasting of the beans or to their fat having been extracted by a petroleum ether of very high boiling point. Cf. also table 12.

41 Zeitschr. Analyt. Chem. B. 21. p. 394.

42 Correspondence of the Association of German Chocolate Manufacturers.

43 Zeitschrift für angew. Chem. 1898, p. 116.

44 We are indebted for this table to the kindness of Dr. Fritsche, Superintendent Meat Inspector at Cleves (Cf. also table of experiments of Matthes & Müller, loc. cit p. — et seq.).

45 Benedikt-Ulzer, Analyse der Fett-und Wachsarten. 5th. edition. 1908. p. 840. also Literature.

46 These high percentages of acid may also be caused by the high percentage of benzine used in the production.

47 A. Ruffin, Pharmaceutische Rundschau 1899, No. 51, p. 820.

48 Therapeutische Monatshefte. 1895. p. 345 and following pages.

49 Compt. rendus de l'acad. des sciences de Paris, Vol. 123,

p. 587.

50 Apotheker-Zeitung 1892, p. 469 and Deutsche Vierteljahrsschrift für öffentl. Gesundheitspflege 1893, No. 3.

51 Pharmaceut. Zeitung 1898, p. 389.

52 Hilger and Lazarus, Compare also Schweitzer, Pharmaceut. Zeitg. 1898, p. 389.

53 Ann. d. Chem. and Pharm. 1841, Vol. 41, p. 125.

54 Ibid. Bd. 118, pag. 151.

55 Berliner Chemische Berichte 1897, pag. 1839.

56 Archiv f. experiment. Pathol. u. Pharmacol. 1895, Vol. 35, pag. 449.

57 Ibid. 1896, Vol. 30, pag. 53.

58 Ibid. 1896, Vol. 36, pag. 66.

59 Ibid. 1888, Vol. 24, p. 101.

60 Therapeut. Monatshefte 1890, p. 10.

61 Semaine médicale 1893, p. 366.

62 Pharmaceut. Centralhalle 1898, p. 901.

63 Dekker (Swiss Weekly Journal, Chem. a. Pharm.) 40, p. 436, 441, 451 u. 463 gives the following figures at 15 ° C.: Water 1800 parts, spirits 1600, pure alcohol 3570, chloroform 3845, ether 25000, acetic unit 3845, benzol 100000 and amylic alcohol 1250.

64 See before.

65 Journal de Pharmacie et de Chimie 1898, p. 176.

66 Ibidem 1897, p. 329.

67 Zeitschrift für analytische Chemie, Vol. 18, p. 346.

68 Aleuron granules were first microscopically observed by H. Molisch (Grundriß einer Histochemie d. pflanzl. Genßmittel in the cellular tissue of the cacao bean. They are very similar to the starch granules of the bean and contain within them a relatively large globoid lime and magnesium phosphates associated with an organic substance (sugar)

which becomes visible in the form of globules when a section is incinerated.

69 Zeitschrift für physiologische Chemie, Vol. 11, p. 207-232.

70 Hygienische Rundschau. 1900. p. 314 & 315.

71 E. S. Bastin, American Journal of Pharmacy 1894, p. 369.

72 Chemischer technischer Centralanzeiger 1886, No. 53, p. 777.

73 Contributions to the establishment of a rational feeding of ruminants. So-called Weender'sche Beiträge, 1864 Number, p. 48 and also Landwirtsch. Versuchsstationen, Vol. 6, p. 497.

74 Zeitschrift für angewandte Chemie 1896, p. 712 und 749.

75 Zeitschrift für Untersuchung von Nahrungs-und Genußmittel. 1898. p. 3.

76 Zeitschrift für öffentliche Chemie 1900, p. 223.

77 Pharmaceutische Zeitung 1898, p. 390.

78 Archiv der Pharmacie 1860, Vol. 153, p. 59.

79 Beitrag zur pharmak. und chem. Kenntnis des Cacaos. Inaug.-Dissertation Dorpat 1375.

80 Untersuchungen über Kakao und dessen Präparate, 1887.

81 Jahresbericht über die Fortschritte der Pharmacognosie etc. 1883, p. 314.

82 Archive der Pharmacie 1893, Vol. 231, p. 694.

83 Zeitschrift des allgem. öster. Apoth.-Vereins 1898, p. 434.

84 Repert. f. anal. Chemie 1885, Vol. 5, p. 178; cf. also the investigations of Mathes & Müller.

85 Grundriß einer Histochemie der pflanzl. Genußmittel, p. 22.

86 Bulletin de la société chimique Paris 1872, p. 33.

87 Pharmaceut. Zeitung Vol. 24, p. 243.

88 Journ. de Pharm. et de Chim. 1883, Ser. V, Vol. 7, p. 506.

89 König, Die menschlichen Nahrungs-und Genußmittel, Vol. 1, p. 261.

90 Zipperer, Untersuchungen über Cacao und dessen Präparate, p. 55.

91 Zeitschr. für Untersuchung von Nahrungs-u. Genußmitteln 1898, No.

92 Repertorium der analyt. Chemie 1885, Vol. 5, p. 178.

93 Compare Matthes & Müller, Z. U. N. 1906, Vol. 12, p. 90 et seq.

94 Almost a tenth part of the ash of the shells consists of silica.

95 cf. Moeller Mikroskopie der Nahrungs-und Genußmittel. Berlin. 1905. II part Springer p. 412.

96 Ztschr. öffentl. Ch. 1899, p. 27.

97 German patent No. 71, 373, 8th. January 1873.

98 Engl. Patent No. 14624, June 16th. 1897.

99 Pharm. Rundschau 1898, p. 781.

100 Ztschr. für chemische Industrie 1878, p. 303, German Patent No. 2112, Sept. 24th. 1878.

101 Annales de Chimie et de Physique, Vol. 183, p. 423.

102 Zeitschrift für Pferdekunde und Pferdezucht 1888, No. 7. Nowadays cacao shells are often added to fodder.

103 Quoted by Filsinger Zeitschr. f. öffentl. Chemie 1899, p. 27.

104 Communication from the Assoc. German Choc. Manufacturers, 19th. year, No. 7.

105 See Mitscherlich, page 111.

106 Practical Guide to Chocolate Manufacture (no date given).

107 Comptes rendus de l'Exposition, quoted by B. de la Roque.

108 Gordian, A., German Chocolate and Sugar Industries,

Vol. 1, p. 22.

109 Correspondence of the Association of German Chocolate Manufacturers 1878, p. 17.

110 Correspondence of Ass. German Chocolate Manufacturers 1891, No. 5.

111 Ibid 1891, No. 7.

112 Zeitschrift für öffentliche Chemie 1898, p. 810.

113 The determining of the fibre is reached by the Weender method.

114 For that purpose boxes with handles and having a capacity of from 10½ to 60 litres are employed, as well as the portable troughs previously mentioned. The transport of the chocolate mass also takes place in boxes made of compressed steel plates (Siemens-Martin), galvanised or otherwise, e. g. as manufactured by the Stamp and Press Works at Brackwede near Bielefeld. The firm of A. Reiche and others also make similar boxes.

115 Muspratt Encyclop. Handbuch der techn. Chemie. Vol. IV, p. 190, 1902.

116 This description is taken from Muspratt, Encycl. Handb. d. Techn. Chemie, Vol. IV. p. 1808 and Mitscherlich: Der Kakao u. die Schokolade p. 115.

117 Constructed by A. Reiche, Sheet Iron Works in Dresden-Plauen.

118 German patent No. 62784.

119 Villon-Guichard, Dictionnaire de Chimie industrielle, Vol. 1 Chocolat.

120 Should such rooms eventually be insulated, the best material for this operation are "Corkstone Plates", as manufactured by various firms (e. g. Korkstein-Werke Coswig i. Sa., etc.).

121 This extensive employment of cacao butter in the preparation of covering material on the one hand, and on the other the consequently increased cost of chocolates rich in fat, have hitherto proved the chief objection to the preparation

of cocoa powder deficient in fatty contents, which we shall discuss later.

122 D.R.P. No. 66606.

123 D.R.P. 74260 of Sept. 3rd. 1893.

124 D.R.P. No. 178897, of July 15th, 1904 (reg. 15th Nov. 1908).

125 This however, is true only to a certain degree, comp. Neumann, The Use of Cacao as a Food Preparation, Munich & Berlin 1906, pag. 97 ff.

126 cf. Z.U.N.G. 1900, vol. 18 p. 171.

127 See enactments of the 16.9.1907 and 10.11.1909 (Coburg): Notices of the Association of German Chocolate Makers XXX, No. 1 21.9.1909, pag. 1.

128 Cf. Z.U.N.G., Bd. 18 Nos. 1 and 2 (1909) p. 178.

129 Eng. Patent No. 20436, 24. 11. 1891.

130 The potash now generally in use is prepared from the carbon of residuary molasses, and is technically considered, very pure. It is supplied by Dr. Hensel & Co., Blumenthal (Hanover).

131 The special model of the Universal Mixer and Kneader has for this purpose (apart from the metal lid shutting down air-tight) a steam drain pipe, which is fitted with a ventilator and led into the open, so that the vapours and chemical exhalations can escape without causing any damage.

132 German Patent No. 30 894. See also Chemiker-Zeitung 1886, p. 1431.

133 Cf. R. O. Neumann, loc. cit. page 98 and following pages.

134 Beckurts Pharmac. Jahresbericht 1883-84, p. 990.

135 The "Machines for packing en masse" Co. Ltd. Berlin, have recently strongly recommended their "wrapping machines, for centres of any shape or consistency, which work automatically, that is to say, it is only necessary to heap the centres in continuous succession in the machine, when they are urged forward and wrapped in paper or other

materials, being finally despatched out of the machine automatically. The wrappers may be simple or double, loose, or tight fitting." Their employment in the packing of chocolate tablets is especially recommended.—And so the problem would be solved! Unfortunately I am in want of personal guidance, never yet having seen the machines in working order, and so not being able to submit any opinion as to their efficiency. Even if they are really able to deal with larger tablets, yet the more critical problem regards the smaller goods, especially in connection with the wrapping in tin-foil.

136 Flour can be more easily blended than starch with the cacao mass, as the granules of starch are only with difficulty crushed.

137 Recently in some inferior kinds of cocoa powder a quantity of oatmeal has often been added (up to 5 percent), causing the preparation to thicken when it is boiled with water.

138 Still better, as less productive of dust, there being a less rapid circulation of air, and also not so wasteful, are the dismembrators as built by Paul Franke & Co.

139 Chemiker-Zeitung 1899. Repert. No. 38, p. 372.

140 Chemiker-Zeitung 1889, p. 408.

141 Beckurts Annual Report of Pharmaceutical Progress etc. 1888, p. 307.

142 See Möller p. 114.

143 Die Nahrungs-und Genußmittel aus dem Pflanzenreiche p. 140.

144 This consists of 15 parts of defatted cacao, 200 parts of arrowroot 50 parts of salep and fifty parts of vanilla-sugar.

145 Krupps Iron Works supply the latest constructions, strongly to be recommended.

146 Arbeiten des kaiserl. Gesundheitsamtes Vol. 15 p. 1-113 and Zeits. f. d. Untersuch. von Nahrungs-und Genußmitteln Vol. 3 21.-25. January.

147 Der Tropenpflanzer 1898, p. 24.

148 Journal of the Society of Arts 1897, Vol. 46, p. 39-40.

149 Compare Gieseler, Vanillevergiftungen, Bonn 1896; Arning (Deutsch. med. Wochenschrift 1897, pag. 435) and Guerin (Annales d'occulistique, 1895 4. October).

150 Arbeiten aus dem Kaiserl. Gesundheitsamte 1899.

151 Journal of the American Chem. Society 1899, Vol. 21, p. 719 and Chem. Ztg. Rep. 1899, p. 275.

152 Berichte der Deutschen Chemischen Gesellschaft Vol. VII, p. 698 and Friedländer, Fortsch. der Theefarbenfabrikation, Berlin 1888, p. 583 and elsewhere.

153 L'état actuel de l'industrie de la parfumerie en France. Revue Générale des sciences pures et appliquées, Paris 1897, p. 663.

154 Chem. Zeit. Repert. 1898, p. 181.

155 Pharm. Zeit. 1888, p. 634 and Pharm. Centralhalle 1898, p. 673.

156 Zeitschr. für angewandte Chemie 1899, p. 428.

157 Pharmaceutische Centralhalle 1898, p. 357.

158 Berlin 1899, Jul. Sprenger, page 53 et seq.

159 K. Dieterich, Die Analyse der Harze, Balsame und Gummiharze, Berlin 1900, page 76.

160 Regulation of 22 and January 1896.

161 See also Farbenzeitung 1909, vol. XV, pages 301, 348, 392 and 436.

162 Ztschr. öffentl. Ch. 1900, page 324, 325.

163 Ztschr. öffentl. Ch. 1900, p. 478.

164 Journ. de Pharm. et Chim. 1898, Vol. 2, page 7.

165 See also Farnsteiner Z. U. N. & G., vol. 23 (1907), page 308.

166 See Farnsteiner's method, Z.U.N. & G., Vol. 13 (1907), page 308.

167 6th. edition, 2nd vol., page 644.

168 Compare: Froehner & Lührig, Z.U.N. & G. IX (1903), p.

257 and Lührig ibid. IX p. 263.

169 cf. the methods of Farnsteiner Z.U.N. & G. XIII, 1907 p. 308.

170 cf. also Farnsteiner Z.U.N. & G. XVI 1908, p. 642 yet according to information from Dr. Böhme from the laboratory of Stollwerk Bros, bluing from red or violet litmus paper should also take place in the case of cacao prepared with potash, and on the contrary the Kurkuma brown not result.

171 Ztschr. für öffentl. Chemie 1900, page 304.

172 Ztschr. für öffentl. Chemie 1900, page 481.

173 Ibid. 1900, pages 86 et seq.

174 Arbeiten aus dem Kaiserl. Gesundh.-Amt 1904, page 20.

175 Ztschr. f. öffentl. Chemie 1907, page 308.

176 Forschungsberichte über Lebensmittel etc. 1896, III page 275, also Beckurt's Jahresbericht der Pharmazie 1896, page 746.

177 Ztschr. f. anal. Ch. vol. 3, page 233.

178 Ztschr. f. anal. Ch., vol. 19, page 246.

179 Journal of Society for Chem. Research 1899, page 556.

180 The solubility of caffeine in carbon tetrachloride is said by Eminger to be 1:100, but Scherr maintains that a much larger quantity is required.

181 Merck's Catalogue of Reacting Agents (2nd. Edition, page 88) gives a convenient method of determining the presence of theobromine and caffeine (Gerard's reaction). We annex an extract.

Gerard's Reaction on Theobromine.

A mixture of 0·05 g of theobromine, 3 ccm of water and ccm of soda wash is decomposed with 1 ccm of a silver nitrate solution 10 percent strong, heated to 60 C. and the solution so obtained cooled down. It then gelatinises very perceptibly. Caffeine does not give this reaction.

Cf. Pharmaceutical and Chemical Journal 1906, p. 476.

Apoth.-Ztg. 1906, p. 432. Pharm. Ztg. 1906, p. 512. Chemical Leaflet 1906 II, p. 167 among others.

182 Soxhlet's so-called steam digester, as constructed by Esser of Munich.

183 Ztschr. f. anal. Ch. 1882, Vol. 22, page 448.

184 Giornale di Farmacia, di Chimica etc. 1898.

185 Lectures for the Establishment of Rational Feeding of Animals (Weender, Lectures), vol. 1864, p. 48. Cf. also "Landwirtschaftl. Versuchsstationen", vol. 4, page 497.

186 Journal of Applied Chemistry 1896, p. 712 & 749.

187 A new process for the determination of crude fibre in food stuffs. Z.U.N. u. G. 1898, p. 3.

188 Ztschr. öff. Chemie 1899, vol. 2, p. 29.

189 Ibid. 1899, vol. 32, p. 479.

190

B. Fischer & Grünhagen, Z. U. N. u. G. 1902, V, p. 83.
P. Drawe, Ztschr. öff. Ch. 1903, IX, p. 161.
G. Lagerheim, Z. U. N. u. G. 1902, V, p. 83.
J. Decker, Schweiz. Wchschr. f. Chem. u. Pharm. 1908, 40, p. 463.
H. Lührig, Bericht d. chem. Unters.-Amtes Chemnitz 1905.

191 Pharmaceutische Zeitung 1889, p. 847.

192 Ztschr. f. öffentl. Chem. 1898, vol. IV, p. 224 u. 225.

193 Untersuchungen über Kakao und dessen Präparate, page 48.

194

See A. Leys, Journ. Pharm. et Chim. 1902 (6), 16, p. 471.
A. Steimann, Ztschr. öffentl. Ch. 1903, 9, p. 239 u. 261.
P. Welmanns, ibid. 1903, 9, p. 93 u. 115.
R. Woy, Schweiz. Wochenschr. f. Chem. u. Pharm. 1903, 41, p. 27.
A. Steimann, ibid. 1903, 41, p. 65.
Fr. David Söhne, Ztschr. öffentl. Ch. 1904, 10, p. 7.
H. Lührig, Bericht d. chem. Unters.-Amtes zu Chemnitz, 1905,

p. 43.

F. Bordas & Touplain, Compt. rendues 1905, 140, p. 1098.

195 Ztschr. f. analyt. Chemie, vol. 22, p. 366.

196 Journal de Pharmacie et Chémie 1877, page 29.

197 Z.U.N. u. G. 1904, 7, p. 471.

198 Ibid. 1909, 18, p. 16 et seq.

199 Ibid. p. 17.

200 Z. U. N. and G. 1909, XVIII p. 19.

201 A word about the R.-M. number seems not out of place here. Baier indeed gives it as an average 1·0 but it varies considerably, as his own investigations show (8 tests of pressed or extracted fats), where there are fluctuations of 1·65 —2·37. Information kindly volunteered by Prof. Härtel and our own experience convinces us that such fluctuations proceed generally from the Glycerine employed, which has itself a R.-M. number, sometimes even amounting to 1·0. It is therefore necessary to fix the standard of Glycerine used in the experiment, only too much neglected in professional investigations.

202 Loc. cit. p. 21.

203 As starting point it may be taken for granted that the R. M. number for milk chocolate is at a minimum 3·75, for cream chocolate 5·5 assuming that 10% cream possesses the R. M. number 3·0 and 20% that between 5·9-6. Various roundabout calculations are so avoided, when the percentages of cream are thus immediately converted into the R. M. number, and the method is quite adequate for estimating purposes.

204 Method of Laxa-Baier, compare Z. U. N. and G. 1909, XVIII p. 18 and 19.

205 Compare: Welmans Zeitschrift für öffentl. Chemie 1900, page 480.

206 The reader who would further consider the form elements of cacao is referred to the excellent paper by Py in the Journal de Pharm. et Chimie 1895. Vol. 1, page 593.

207 Compare: E. Guenez, Revue internationale des

falsifications des denrées alimentaires 1895. Vol. 9, pages 83-84.

208 Chemiker-Zeitung 1890. Vol. 14, Rep. page 48.

209 Zeitschrift für öffentliche Chemie 1900, page 480.

210 Cf. Beytheon, Pharm. Central-Halle 47, page 749.

211 Compare page 283 and the remarks there.

212 There may be, however, an enormous difference.

213 Report and stenogr. prot. publ. by the periodical Nahrungsmittel-Untersuchung u. Hygiene; Pertes, Wien, page 60.

214 Comp. Dr. Böhme, The Chocolate and Confectionery Industries, VI 1911, No. 37. The assembly came to an agreement on all points discussed, and it would be well to repeat the resolutions here.

215 Dissimilar to all other existing definitions and adapted to the new method with slightly roasted beans only.

216 I. e. about 2·3-2·5 kilos of potash to 100 kilos of cacao mass.

217 Thus satisfying the demands of the Free Association of German Food Chemists.

218 Would thus be too little according to the regulations under II.

219 Cocoa powder may thus, according to international custom, also be flavoured with spices.

220 Cf. in this connection page 204 and tables 19 & 20.

221 According to recent resolutions of the Free Union (cf. page 282) the percentage of sugar in chocolate (together with additions for medicinal and dietetic purposes) may not exceed a total 68%; but there is no fixed standard for the fatty contents, except in the case of milk chocolates etc.

222 The excessive use of cacao butter as an admixture has lately assumed large proportions. In commerce there are to be found many preparations designated as "pure cacao and sugar" which contain only 15 or 20% of cacao with 50% of fat,

which are said to met a need of the public, but the maintenance will scarcely hold water.

223 The Roumanian law admits of the sale of a cacao prepared from the unshelled bean and only precludes secondary admixtures of shell.

224 Better albumose, or still better not included at all, as this conversion of the albumen is by no means proved.

225 Accordingly an addition of cacao butter would be objectionable. But with 70% of sugar, admixture of cacao butter is unconditionally necessary, where by the pure cacao material sinks to between 10% and 20%.

226 Editor's note: These figures are subject to correction, as they do not tally with the majority of accepted results.

227 Cf. note on page 294 under 2.

228 Whilst in Germany such admixture is not permissible at all.

229 Editor's note: These values would seem to require some revision, as generally only the very inferior cacaos, like St. Thomé, Domingo, Cuba and Haiti, show a lower ash percentage than 3·5%; Ariba, Porto Cabello, Caracas and Guayaquil cacaos show a higher percentage the same remark applies also to the fibre content.

230 This also requires revision, as on boiling 7·5 grammes cacao with 250 grammes water there will always be a sediment after the solution has stood for some minutes.

231 Requiring revision Cf. remarks on previous page and also the values of raw fibre found by Filsinger. Editor's note.

232 Requires revision, compare page 261. Editor's note.

233 We would prefer Eminger's method. — Editor's note.

234 Cf. above, § 2, 1 and 2.

235 The "Deutsche Nahrungsmittelbuch" issued by the Association of Manufacturers and Dealers Trading in Articles of Consumption has unfortunately only complexed matters as it was a private undertaking and has endeavoured to sanction various usages, better termed misusages, such as the use of

forbidden preserving and conserving agents, artificial colouring stuffs etc. It is true that the part connected with cacao preparations constitutes a glorious exception, and also that there are recent indications of an agitation to reform the whole code.

236 Both are designs of the firm J. M. Lehmann, by whom they have been obligingly placed at our disposal.

237 Hahn-Holfert, Spezialitäten und Geheimmittel, page 300.

238 Pharmazeutische Zeitung 1888, page 512.

239 German patent No. 182747 (Jan. 4th 1905) 182748 (May 4th 1906).

240 German patent No. 189733 (26th February 1906), 189734 (Dec. 11th 1906).

241 Which would seem to be the only proper employment of the total patent claim.

242 According to Dieterich (Neues Pharmazeutisches Manual, 7. edition page 191) prepared barley meal is obtained as follows: 1 kilo barley flour is firmly pressed into a suitable metallic (tin) vessel, so that it is about 2/3 full and then heated on a water bath for 30 hours in all. After the lapse of 10 hours the powder is removed and ground in a mixer them again placed in the vessel and re-heated for 10 hours. After twice repeating this manipulation, about 900 grammes of a reddish mass will be obtained which is prepared barley meal.

243 Apotheker-Zeitung 1900, page 181.

244 Compare Aufrecht, Pharm. Zeitung 1910, page 558.

245 The absurdity of this process is too evident to need remark; would it not have been better, if the process had not had the sanction of the patent mark? The treatment, which the cacao here undergoes, is so barbarous, that the product must always be spoiled. The only point attained is the complete gelatinisation of the starch, which by further heating is to some extent converted into dextrin. Caramelizing cannot and will not take place by heating gelatinised starch in mixtures with a dry substance, as it occurs in cacao. But in addition, the claim is weak that cacao so mistreated would be especially

suitable for diabetics, since cacao serves that purpose a great deal better. The addition of albumin every properly disintegrated is not at all new, for mixtures of albumin and cacao have existed for a very long time.—Editor's note.

246 Instead of which pure milk powder may also be used.

247 All cacao preparations, to which albumin is added, require a large amount of cacao butter as the albuminoids largely absorb the fat.

248 The composition of the preparation must be stated on the wrapper as such terms as "Natur-cacao" and "Natur-chocolate" are liable to lead the purchaser astray.—Editor's note.

249 Alfr. Beddies, Ueber Kakaoernährung, Berlin 1897.

250 Plasmon is an albuminoid preparation from milk, to which a little sodium bicarbonate is added to effect complete solution.

251 Somatose is a nutritive preparation made from meat and contains the nitrogenous constituents of the muscle flesh exclusively in the form of an easily soluble albumose.

252 Tropon is a mixture of 2 parts flesh albumin (from muscle flesh and fish) and one part plant albumin.

253 The preparation must also bear on the wrapper a statement of its composition in order not to mislead the purchaser.

German Text to Fig 5.

Ausgestellt vom Verband deutscher Chocoladefabrikanten.

Sitz Dresden

Verbrauch von Rohkakao
1896-1901

in *Frankreich. Grossbrittanien. Holland*
den *Verein. Staaten v. N-A.* und *Deutschland*
in 1000 Dz. (100 kg).

Einfuhr von Rohkakao über die Deutsche Zollgrenze
1883-1901
in Doppelzentnern.

Prozentuale-Steigerung
des durchschnittl. Verbrauchs
von *Kakao* (in Bohnen) *Kaffee* u. *Tee*
in *Deutschland* verglichen mit dem
Stande von 1840.

ANTONREICHE A. G.
:DRESDEN:

Manufacturer of Chocolate Moulds, decorated tin Boxes etc.

ESTABLISHED 1870

Chocolate Moulds of every description latest are

"Plattinol" Moulds which impart a **rich lustre** and **finish** to the chocolate

Chocolate Drop Presses for Paste Chocolate for hand and for liquid chocolate, Automatic Power

Chocolate Covering Apparatus

Machine for granulated Chocolate (Streussel-Machine)

Decorated Tin Boxes

WRITE FOR CATALOGUES AND PRICES
About 2000 employees

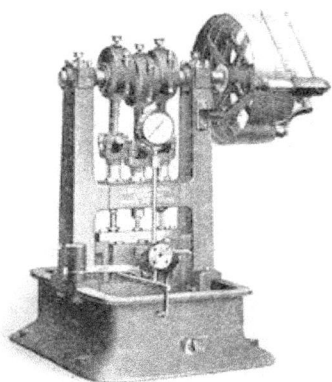

J. M. LEHMANN ∘ DRESDEN
Founded 1834

Oldest and largest Engineering Works for the construction of modern Machines for the Manufacture of Cocoa and Chocolate

PARIS, 1, Passage St. Pierre Amelot.
NEW YORK, 13/15, Laight Street.

Sole Agents for Great Britain>: **Bramigk & Co., London E,** 5, Aldgate

Melangeurs of latest construction

Capacities from ½ to 6 Cwt.

With automatic discharge, saving Time and Labour.

Easy handling and economical working

Refining Machines

with 3, 4, 5, 6 and 9 rollers of granite or chilled metal (steel) with water-cooling

Very large output, great saving of space and driving power. Extraordinary Fineness of the finished material

J. M. LEHMANN ○ DRESDEN
Founded 1834

Oldest and largest Engineering Works for the construction of modern Machines for the Manufacture of Cocoa and Chocolate

PARIS, 1, Passage St. Pierre Amelot.
NEW YORK, 13/15, Laight Street.

521

Sole Agents for Great Britain: **Bramigk & Co., London E**, 5, Aldgate

Chocolate Cooling Plants
improved construction

Mechanical Cooling Plant in conjunction with Tempering and Moulding Machines

Melting Pan, automatic Tempering Machine, one or more Moulding Machines, Shaking Tables and continuously working Cooling Chamber with forced air circulation

Largest output. Great Saving of time and Labour. Automatic conveyance of the full moulds over the shaking table and through the cooling chamber to the packing room, and conveyance of the empty moulds back to the moulding machine

Kunstanstalt vorm.
ETZOLD & KIESSLING A.-G.

CRIMMITSCHAU, SAXONY

The Chromolithographic Institute

Patent Folding & Fancy Paper Boxes of all kinds, for commercial and other purposes, Showcards, Labels, Wrappers etc., Calendars, Catalogue Covers, Reproduction of articles of merchandise in actual colours, Insets and Advertising Novelties

Specialists in Chocolate Wrappers and Boxes

Cacao Pulveriser

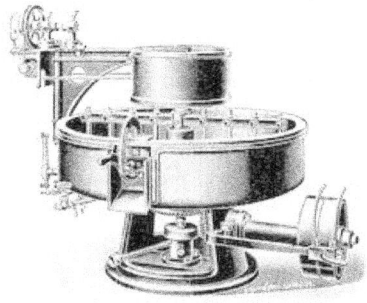

Chocolate Tempering Machine

All machines

for the

manufacture of Chocolate, Cocoa and Confectionery

Paul Franke & Co.

Engineering Works

Leipzig-Böhlitz-Ehrenberg

Catalogues and Estimates
on demand

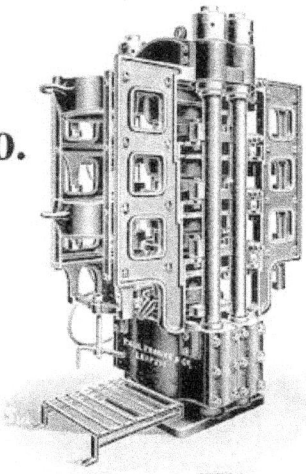

Strong Hydraulic Cocoa Press

All machines for the manufacture of Chocolate, Cocoa and Confectionery

525

Paul Franke & Co.

Engineering Works

Leipzig-Böhlitz-Ehrenberg

Catalogues and Estimates on demand

M . KRAYN, Verlagsbuchhandlung, BERLIN W. 10

In meinem Verlage erschienen:

Die Chemie in industrie, Handwerk und Gewerbe von **Joseph Spennrath**, weil. Direktor der gewerblichen Schulen der Stadt Aachen. *Fünfte vermehrte und verbesserte Auflage*, bearbeitet von **Dr. L. Sender**. Ein Lehrbuch zum Gebrauch an Schulen, sowie zum Selbstunterricht. Preis brosch. **Mk. 3.60**, kart. **Mk. 3.90**.

Die Bedienung und Wartung elektrischer Anlagen und Maschinen von **Joseph Spennrath**, weil. Direktor der städt. gewerbl. Schulen und der Kgl. Baugewerbeschule in Aachen. *Zweite, vollständig neu bearbeitete u. bedeutend erweiterte Auflage* v. Dipl.-Ing. **Franz Menge**. I. E i n f ü h r u n g   i n   d i e   G r u n d l a g e n   d e r   E l e k t r o t e c h n i k . Mit 207 Abbildungen und 1 Tafel. II. E i n f ü h r u n g   i n   d e n   B a u   u n d   d i e   W i r k u n g s w e i s e   d e r   S t r o m e r z e u g e r . Mit 210 Abbildungen. Preis pro Band brosch. **Mk. 2.80**, kart. **Mk. 3.25**. Preis komplett I./II. brosch. **Mk. 5.50**, kart. **Mk. 6.—**.

Temperaturmeßmethoden. Handbuch zum Gebrauch bei praktischen Temperaturmessungen von **Bruno Thieme**. 35 Figuren im Text. Preis brosch. **Mk. 4.—**[, geb. **Mk. 5. —**.

Rechenhilfsbuch. Berechnungstabellen für Handel und

526

Industrie, insbesondere für jede Lohn-und Akkordberechnung, nach langjähriger Erfahrung herausgegeben von **G. Schuchardt**. D. R. G. M. *Dritte verbesserte Auflage*. Preis geb. **Mk. 5.—**. F ü r   g r ö ß e r e   B e t r i e b e   u n e n t b e h r l i c h ! Durch eine ganz neuartige, geschützte Register-Anordnung vermittelt das Schuchardt'sche Rechen-Hilfsbuch   s c h n e l l s t e   A u ffi n d u n g der gewünschten Zahlen.

Der Praktische Lohnrechner. Handbuch für jede Lohnberechnung von **G. Schuchardt**. Preis geb. **Mk. 2.** —. Es sind in diesem Buche die Lohnsätze von 7½-75 Pf. in Intervallen von 2½ Pf. aufgenommen, ferner auch die häufig üblichen Lohnsätze von 18, 22, 28, 32 Pf. Die Stundeneinteilung ergibt die Uebersicht von ¼ bis 99¾ Stunden. F ü r   k l e i n e r e   u n d   m i t t l e r e   B e t r i e b e   u n e n t b e h r l i c h !

Die Kontrolle industrieller Betriebe. Praktische Anleitung zur Durchführung einer modernen Betriebskontrolle von **G. Schuchardt**. Preis brosch. **Mk. 1.60.**

Der praktische Maschinenwärter. Anleitung für Maschinisten und Heizer sowie zum Unterricht in technischen Schulen von **Paul Brauser**, Oberingenieur des Dampfkessel-Revisions-Vereins für den Regierungsbezirk Aachen und **Joseph Spennrath**, weil. Direktor der gewerblichen Schulen der Stadt Aachen. *Vierte verbesserte und vermehrte Auflage*. Mit 42 Holzschnitten. Preis kart. **Mk. 1.50.**

Der praktische Heizer und Kesselwärter von **Paul Brauser**, Oberingenieur des Dampfkessel-Revisions-Vereins für den Regierungsbezirk Aachen und **Joseph Spennrath**, weil. Direktor der gewerblichen Schulen der Stadt Aachen. *Siebente verbesserte Auflage mit 60*

Holzschnitten. Preis kart. **Mk. 1.80.**

Zu beziehen durch jede Buchhandlung oder direkt vom Verlag

M . K R A Y N, Verlagsbuchhandlung, B E R L I N W. 10

<u>Für alle Kalkulationsbüros!</u>

Rechen-Resultate

Tabellen zum Ablesen der Resultate von Multiplikationen und Divisionen (in Bruchteilen und ganzen Zahlen)

von 1 bis 1000

Zum praktischen Gebrauch für Stückzahl-, Lohn-und Prozentberechnungen, sowie für jede Art Kalkulation

Preis gebunden 10 Mark

Herausgegeben von
F. TRIEBEL, Kaiserlicher Revisor der Reichsdruckerei

Die Papierverarbeitung
von MAX SCHUBERT
weiland Fabrikdirektor a. D., Prof. a. d. Königl. techn. Hochschule zu Dresden

I. B a n d:

Die Kartonnagen-Industrie

Praktisches Handbuch für Techniker, Kartonnagen-Fabrikanten und Buchbinder

Mit 479 Illustrationen und 2 Musterbeilagen

Preis broschiert 10.—Mark, gebunden 11.50 Mark

II. B a n d:

Die Buntpapier-, Tapeten-, Briefumschlag-, Düten-oder Papiersack-, Papierwäsche-und photographische Papier-

Fabrikation

Praktisches Handbuch für Techniker, Buntpapier-, Tapeten-und Dütenfabrikanten-Direktoren

Mit 278 Illustrationen

Preis broschiert 10.—Mark, gebunden 11.50 Mark

I. u. II. Band, zusammen bezog., brosch. 18.—M., geb. 20 —M.

Ausführlicher Prospekt gratis

Zu beziehen durch jede Buchhandlung oder direkt vom Verlag